HORTI-BUSINESS MANAGEMENT

HORTI-BUSINESS MANAGEMENT

Chirag Sah

2025

Agri Horti Press

New Delhi-110 084

ISBN: 978-93-9296-931-7 (Hardbound)

Published By

Agri Horti Press

New Delhi

E-mail: agrihortipress@gmail.com

Preface

"Horti-Business Management" represents a comprehensive exploration of the principles, practices, and strategies essential for successful management in the horticultural industry. This book is designed to serve as a practical guide and valuable resource for horticultural professionals, entrepreneurs, students, and enthusiasts seeking to navigate the complexities of horticultural business operations.

The horticultural industry encompasses a wide array of businesses, including nurseries, greenhouse operations, landscaping companies, floriculture enterprises, and specialty crop production facilities. In today's rapidly changing economic landscape, effective management is essential for the sustainability and success of horticultural businesses, regardless of their size or scope.

The primary objective of this book is to provide readers with a comprehensive understanding of horticultural business management, covering key topics such as business planning, financial management, marketing strategies, human resource management, and sustainable practices. Drawing upon the expertise of industry professionals and scholars, each chapter offers practical insights, case studies, and best practices to help readers develop the skills and knowledge necessary to thrive in the horticultural marketplace.

Chapters in this book address a range of critical issues facing horticultural businesses, including market analysis, product development, supply chain management, regulatory compliance, and risk assessment. Additionally, special emphasis is placed on emerging trends and opportunities in the horticultural sector, such as urban agriculture, vertical farming, agri-tourism, and e-commerce.

As we delve into the complexities of horticultural business management, it is important to recognize the unique challenges and opportunities inherent in the industry. From fluctuating market demands to seasonal variations in production, horticultural businesses must adapt and innovate to remain competitive in today's dynamic marketplace. By embracing sound management practices, leveraging technological advancements, and fostering a culture of sustainability and innovation, horticultural businesses can position themselves for long-term success and growth.

We extend our sincere gratitude to the contributors whose expertise and dedication have enriched this book. It is our hope that "Horti-Business Management" will serve as a valuable resource and practical guide for horticultural professionals and aspiring entrepreneurs alike, empowering them to navigate the complexities of the horticultural industry with confidence and competence.

Chirag Sah

Contents

1

HORTI-BUSINESS: OVERVIEW

Horticulture is the art of cultivating plants in gardens to produce food and medicinal ingredients, or for comfort and ornamental purposes. Horticulturists are agriculturists who grow flowers, fruits and nuts, vegetables and herbs, as well as ornamental trees.

The study and practice of horticulture have been traced back thousands of years. Horticulture contributed to the transition from nomadic human communities to sedentary, or semi-sedentary, horticultural communities. Horticulture is divided into several categories which focus on the cultivation and processing of different types of plants and food items for specific purposes. In order to conserve the science of horticulture, multiple organizations worldwide educate, encourage, and promote the advancement of horticulture. Some notable horticulturists include Luca Ghini and Luther Burbank.

Horticulture is an important part of a nation's economy. In India, Horticultural crops are important to the economy because they provide jobs, provide raw materials for food processing businesses, and generate revenue owing to increased output and profitability. Furthermore, Foreign exchange profits a lot from exports of horticulture products. In this article, we will read about Types of Horticulture Farming in India.

The term Horticulture is derived from the Latin words hortus (garden) and culture (cultivation). It is the branch of agriculture. In other words, it is the cultivation, production, and sale of vegetables, fruits, flowers, herbs, ornamental or exotic plants.

Horticulture farming also aims to enhance the quality of life, as well as the beauty, sustainability, and recovery of our ecosystem and the human condition.

Plants, crops, and green spaces help to sustain and enrich our lives by providing healthy food, beautifying our homes and communities, and lowering our carbon footprint.

Horticulture is a specific part of agriculture that deals with many techno-commercial aspects and plant cultivation. Horticulture business demands intensive management and a significant investment of time and energy. It is important to be aware of your motivation for establishing a horticultural enterprise.

Definition and types

Horticulture involves plant propagation and cultivation to improve plant growth, yields, quality, nutritional value, and resistance to insects, diseases, and environmental stresses. It also includes plant conservation, landscape restoration, soil management, landscape and garden design, construction and maintenance, and arboriculture. The word horticulture is modeled after agriculture; it derives from the Latin words hortus and cultura, which mean "garden" and "cultivation", respectively. In contrast to agriculture, horticulture does not include intensive crop farming and large-scale crop production or animal husbandry. Additionally, horticulture focuses on the use of small plots with a wide variety of mixed crops while agriculture focuses on one large primary crop at a time.

Types of horticulture

There are several major areas of focus within the science of horticulture. They include:

- Olericulture: the production of vegetables.
- Pomology, also called fruticulture: the production of fruits and nuts.
- Viticulture: the production of grapes (largely intended for winemaking).
- Floriculture: the production of flowering and ornamental plants.
- Turf management: the production and maintenance of turf grass for sports, leisure and amenity use.
- Arboriculture: the cultivation and care of individual trees, shrubs, vines, and other perennial woody plants, primarily for landscape and amenity purposes.
- Landscape horticulture: the selection, production and care of plants used in landscape architecture.
- Postharvest physiology: the management of harvested horticultural crops to retard spoilage while stored or transported.

Horticulture Farming in India

Because of India's diverse agro-climatic conditions, all sorts of fresh fruits, vegetables, and medicinal plants can be grown in different regions of the nation. In addition, There is a growing health consciousness among the Indian population.

Since the majority of the population is vegetarian in India. There is a strong demand for fruits and vegetables. However, When compared to the country's current demand for horticulture products, the country's production is far lower. Hence, there is huge scope for farmers and businesses in the Horticulture sector in India.

- Primary objectives of National Horticulture Board to Provide Subsidy For Horticulture Business

- National Horticulture Board (NHB) was set up by the Government of India in 1984. Basically, it is an autonomous society under the Societies Registration Act 1860.
- The objectives of the National Horticulture Board are the development of hi-tech commercial horticulture, development of modern post-harvest management infrastructure, promotion, market development of fresh horticulture produce, and more.
- They also carry out studies and surveys to identify constraints. Additionally, they develop short and long-term strategies for the systematic development of horticulture.
- Also, the agency provides technical services including advisory and consultancy services. National Horticulture Board in India provides several subsidies for different horticulture initiatives.

Different Schemes of Subsidy For Horticulture Business

(A) Commercial Horticulture Scheme

1. Commercial Horticulture through Production and Post Harvest Management

Credit linked projects relating to the establishment of commercial production units in the open field as well as under protected conditions and projects on Postharvest Management and primary processing of products are eligible for assistance under this scheme.

2. Commercial Horticulture Development in open field conditions

National Horticulture Board accepts integrated commercial horticulture development projects in open field conditions on project mode, including components viz planting material, plantation, irrigation, fertilization, mechanization, precision farming, GAP, etc. for projects covering an area of 2.00 ha. (5 Acres).

Integration of production unit with on-farm PHM components and primary processing unit shall also be allowed in project mode. Credit linked back-ended subsidy @ 40% of the total project cost limited to Rs 30.00 lakh per project in general areas and @ 50% of project cost limited to Rs. 37.50 lakh in NE Region, Hilly, and Scheduled areas.

3. Commercial Horticulture Development in a protected cover

The Board accepts commercial horticulture development projects under protected cover on project mode including components viz planting material, plantation, irrigation, fertilization, mechanization, etc for projects having an area over 2500 SQ meter.

Credit linked back-ended subsidy @ 50% of the total project cost limited to Rs 56.00 lakh per project as per admissible cost norms for greenhouses, shade net house, plastic tunnel, anti-bird /hail nets & cost of planting material, etc.

4. Integrated Post Harvest Management projects

National Horticulture Board accepts Integrated Post Harvest Management projects relating to Pack House, Ripening Chamber, Refer Van, Retail Outlets, Pre-cooling unit, Primary processing, etc. NHB will also take up projects in component mode and for standalone projects of PHM components.

Credit linked back-ended subsidy @ 35% of the total project cost limited to Rs 50.75 lakh per project in the general area and @ 50 % of project cost limited to Rs. 72.50 lakh per project in NE, Hilly, and Scheduled areas.

(B) Cold Storage Scheme

Credit linked projects relating to Cold Storage including Controlled Atmosphere (CA) and their modernization are eligible for assistance under this component.

Subsidy need not be credit-linked for the institutions like Public Sector Units, Panchayats, cooperatives, registered societies/trust, and public limited companies provided they can meet a remaining share of the project cost out of their own resources.

Such projects will have to be appraised by an appraising agency approved by NHB. The assistance will be given as a subsidy @ 35% of the capital cost of the project in general areas and 50% in the case of NE, Hilly & Scheduled Areas for a storage capacity of 5000 MT up to 10000 MT.

(C) Technology Development and Transfer Scheme

National Horticulture Board accepts projects for the popularization of identified new technologies/tools/techniques for commercialization and adoption through following sub-components of the scheme

1. Setting up of block/mother plant and rootstock nursery
2. Acquisition of technologies including import of planting material from other countries
3. Import/Procurement of Machines and Tools for horticulture for demonstration purpose
4. Development & Transfer of Technology
5. Product Promotion and Market Development Services
6. Exposure visit of farmers (Outside State)
7. Accreditation and Rating of Fruit Plant Nurseries

The financial success of a horticulture business often depends on having adequate finance to fund you until your crop is fully established and generating the profits you planned for. Plan rightfully and contact the nodal officers to get a subsidy for the horticulture business in time.

Types of Horticulture Farming in India

There are numerous types of Horticulture Farming in India. However, the most significant ones are as follows"

Arboriculture

How plants grow and adapt to their environment is studied in arboriculture. In arboriculture, Individual trees, shrubs, vines, and other perennial woody plants are studied. In simple words, it generally relates to maintaining individual woody plants and trees for long-term landscape and amenity purposes.

Turf Management

Turf Management is one of the important types of horticulture farming in India. It describes the work to manage the sports pitches. For example, management of cricket pitch, maintenance of golf pitch etc. comes under Turf Management.

Floriculture

Floriculture is the process of farming flowers. It is a branch of horticulture associated with flower and ornamental crops for use in gardens and floristry. Floriculturists spend most of their time cultivating new varieties of flowers through plant breeding.

In simple words, it is the growing, maintenance, and marketing of ornamental flowers. However, some florists also do the extraction of essential oils, edible pigments.

Landscape Horticulture

In landscape gardening, attractive plants are planted in such a manner as to create a scenic appearance. In simple words, Landscape gardening is an artistic branch of Horticulture.

Landscape Horticulture is a mixture of science and art. It is a vital and emerging sector considering the scope of horticulture farming in India.

Pomology

Pomology is the study of fruit, specifically the science of growing fruit and nuts. Fruits are always on-high demand. Therefore, Pomology has a significant role in horticulture farming of India.

Among a pomologist's responsibilities is the creation of new cultivars of fruits and vegetables. Fruit and nut cultivars are continuously being modified to enhance disease resistance and other characteristics. Moreover, Pomologists also investigate fertiliser and pruning strategies for growing healthy trees.

Postharvest Physiology

Postharvest Physiology is used to determine the best storage and transportation conditions to optimise shelf life after harvest.

As the name suggests, Postharvest Physiology focuses primarily on technology after harvesting. In Postharvest Physiology, we also get to study optimal storage and transportation conditions to increase the shelf life.

Commercial Horticulture Farming

As we all know, People make money from commercial horticulture by cultivating plants for other people.

It requires a thorough understanding of the market, including what customers want, what their likes and preferences are, and what is readily accessible and what is in short supply.

Horticultural organizations

There are various organizations worldwide that focus on promoting and encouraging research and education in all branches of horticultural science; such organizations include the International Society for Horticultural Science and the American Society of Horticultural Science.

In United Kingdom, there are two main horticulture societies. The Ancient Society of York Florists is the oldest horticultural society in the world and was founded in 1768; this organization continues to host four horticultural shows annually in York, UK. Additionally, The Royal Horticultural Society, established in 1804, is a charity in United Kingdom that leads on the encouragement and improvement of the science, art, and practice of horticulture in all its branches. The organization shares the knowledge of horticulture through its community, learning programs, and world-class gardens and shows.

The Chartered Institute of Horticulture (CIoH) is the professional body which represents horticulturists in Great Britain and Ireland while also having an international branch for members outside of these islands. The Australian Society of Horticultural Science was established in 1990 as a professional society to promote and enhance Australian horticultural science and industry. Finally, the New Zealand Horticulture Institute is another known horticultural organization.

In India, Horticultural Society of India (now Indian Academy of Horticultural Sciences) is the oldest society which was established in 1941 at Lyallpur, Punjab (now in Pakistan) but was later shifted to Delhi 1949. The another notable organisation in operation since 2005 is the Society for Promotion of Horticulture based at Bengaluru. Both these societies publish scholarly journals viz., Indian Journal of Horticulture and Journal of Horticultural Sciences for the advancement of horticultural sciences.

The National Junior Horticultural Association (NJHA) was established in 1934 and was the first organization in the world dedicated solely to youth and horticulture. NJHA programs are designed to help young people obtain a basic understanding of horticulture and develop skills in this ever-expanding art and science.

The Global Horticulture Initiative (GlobalHort) fosters partnerships and collective action among different stakeholders in horticulture. This organization has a special focus on horticulture for development (H4D), which involves using horticulture to reduce poverty and improve nutrition worldwide. GlobalHort is organized in a consortium of national and international organizations which collaborate in research, training, and technology-generating activities designed to meet mutually-agreed-upon objectives. GlobalHort is a non-profit organization registered in Belgium.

To Start a Horticulture Business

- The world would be a different place if we did not have the greenery and the plants that surround us. These two are essential and have several functions in making our lives easy.
- They are responsible for purifying our air, feeding us, and even giving us a green and beautiful environment, making the world a better place for us.
- Starting a business in horticulture is one of the best career choices for those that love working outdoors. It, however, requires a number of things such as innovation, proper business management, advanced technology, and a passion for one to come up with fresh produce.
- It also requires one to have the desire to create and look for ways to build a sustainable and better future for the business. The horticulture business can be rewarding depending on what you are doing.
- In this business, you will be involved in different things, such as working with crops, building and designing parks, experimenting with new plants, and scientifically modifying plants to make them better.

However, do you know what you need for you to start a horticulture business and to avoid the reasons for the failure of small businesses?

Greenhouse Structures and Lighting

One of the most important things in this business is the greenhouse and lighting. You will not start a business in horticulture without a greenhouse that has the right tools such as lighting since it forms the backbone of this business.

You will need to think about the structure of the greenhouse that fits the type of plants that you will be growing. The plants will also need proper lighting for them to grow well and to provide good produce.

You will need to compare grow lights and choose the one that best fits the needs of your plants. This is important if you need to be successful in your business.

Types of Plants to Grow

After getting the greenhouse and proper lighting, you will need to come up with a planting schedule for your use. One of the first steps to take when deciding what to grow is to have a look at what people currently grow in your location.

In addition, you will also need to talk to buyers and see what the market needs the most. There are many factors to consider when making a choice on the plants to grow, and you need to ensure that you have put all of them into consideration.

These factors might also include the demand and supply of plants, the market availability in your location, and your market radius, or rather how far you would like to supply your plants. That notwithstanding, you need to be growing the most profitable crops.

Horticulture Experience

- Starting a horticulture business is not a walk in the park. You will need to know a lot of things, some of which cannot be learned in class. You, therefore, need to evaluate if you have the right experience to start growing plants for commercial purposes.
- If you do not, then you might need to first look for a job in a commercial horticultural company to gain the required experience.
- This will make sure that you know everything with regards to the commercial production of plants. This experience cannot be achieved by simply taking courses. With this experience, you will understand what is needed of you and if it is worth the sacrifice.
- You will also know if you have all the time that the business demands.

Horticulture Regulations

- Finally, you need to understand all the regulations governing the horticulture business in your location. This will require you to do enough research since the regulations may vary from one place to another.
- You need to find out about things such as the maximum and minimum measurements required for a greenhouse when it comes to the commercial production of plants.

- These measurements might pose a challenge when getting a greenhouse in some areas, meaning that you might be forced to construct your own.
- You might also need to find out about any plants that might not be allowed in your area. There are areas that do not allow the growing of certain plants for commercial purposes. You can also look at the pesticides and fertilizers that are allowed or not allowed in your location.
- In conclusion, considering the above tips will ensure that you have a profitable horticultural business.

Benefits of Horticulture Farming in India

Horticulture benefits are not limited to the environment only. Rather, Horticulture farming can be a profitable business in the long run. The reason is it provides several benefits. Let us look at some of the most important ones.

- Horticulture is one of India's most important development drivers. It can be more lucrative than agriculture in several cases (food grains mainly).
- Horticulture offers a variety of professions in the primary, secondary, and tertiary sectors, as well as in the government.
- Small and medium farmers benefit from horticultural crops, fruit, and vegetable crops.
- Horticulture also allows people to consume a diversified and balanced food, which is important for a healthy lifestyle.

Exporters of scheduled horticulture products having APEDA Registration can expand their business worldwide and earn profits. Moreover, APEDA Registration allows exporters and farmers to have financial aid and attend various training programs to expand the business.

Horticultural Producers' Cooperative Marketing and Processing Society

Horticultural Producers' Cooperative Marketing and Processing Society, popularly known by its acronym, HOPCOMS, is a farmers' society founded in 1965 for the direct marketing of farm produces. The society is headquartered in Lalbagh, Bengaluru, in the south Indian state of Karnataka and its activities are spread in the districts of Bangalore Urban, Bangalore Rural, Kolar, Ramanagar, Mandya, Mysuru and Chikkaballapura.

The origin of HOPCOMS was in 1959 when Mari Gowda, the then director of the department of horticulture, founded the Bangalore Grape Growers' Marketing and Processing Co-operative Society for promoting grape farming by providing farming know how to the grape farmers and arranging a marketing set up for their products. The society had its operations in Bangalore, Kolar, Mysore, Tumkur, Mandya and Mangalore districts. A few years later, the society started dealing in other farm produces and rebranded itself as the Horticultural Producers' Co-operative Marketing and Processing Society in 1965, registering itself as a cooperative society. It comes under the jurisdiction of the department of horticulture, Government of Karnataka.

The main activity of the society is the maintenance of a marketing system by which the farmers are ensured of a good price for their produces and the consumers are provided with quality goods at affordable prices. The society has eliminated the middle men and acts as the only link between the farmers and the consumers. As of March 2014, the society has a paid up capital of ₹26.59 million composed of Class A (farmers), Class B (NGOs, institutions) and Class C (government) shares and a member strength of 12,680 farmers. The organization is managed by a board of directors, reporting to the department of horticulture of the Government of Karnataka. Eleven of the 15 member board are elected from the farmers while the state government has three appointed officers and a nominee on the board. It employs 650 permanent staff and 790 temporary workers. The President is elected from the Class A members while the Managing Director is an appointee of the state government.

Objectives

- Dissemination of knowledge by way of training, technical advice and printed literature.
- Supply of farm equipment, consumables such as manure and grafts.
- Arranging storage facilities including cold storage.
- Facilitating marketing of the produces through own marketing set up.
- Direct supply of fruits and vegetables to factories, hospitals, hostels, clubs, social functions and processing industries.

Activities

The activities of HOPCOMS are threefold; procurement, storage and distribution. It has opened four procurement centres at Sarjapura, Channapatna, Hosakote and Kanakapura where facilities have been arranged for acceptance, weighment and grading of produces. The payment to the farmers are distributed immediately on acceptance of goods. The goods are then transferred to central godown, processing units or retail outlets as per requirement. The society runs 276 retail outlets of which 238 retail outlets in the urban and rural areas of Bengaluru and 19 outlets in Kolar district.

HOPCOMS has fruit processing units which processes Grape, Mango and Orange juices for the retail market. It also has a distribution system for supplying fruits and vegetables to government hospitals such as Victoria Hospital, Leprosorium Hospital, Bowring & Lady Curzon Hospitals, NIMHANS and Kidwai Memorial Institute of Oncology, factories such as Indian Telephone Industries and Hindustan Machine Tools, hostels and clubs and also undertakes supply of produces for social and marriage functions. The total quantity of goods handled by HOPCOMS in a day is around 100 tonnes. It also supplies equipment, manure and consumables to the farmers at reduced rates and has plans to open a walk-in supermarket and 100 modern outlets with funding from the Karnataka Horticultural Federation. The society organizes periodic fruit and vegetable festivals in various places in Karnataka where special stalls are set up for direct selling of produces. HOPCOMs is also associated with other organizations dealing in fruits and vegetables such as National Agricultural Cooperative Marketing Federation of India (NAFED), Himachal Pradesh Horticultural Produce Marketing and Processing Corporation (HPMC), GROWREP, VEFCO and National Diary Development Board (NDDB).

The society avails financial assistance from government and financial institutions for its developmental activities. It has collected grants from Zilla Parishads and the National Horticultural Board and a loan of ₹44.80 million from the National Cooperative Development Corporation of which ₹21.50 million has been paid back. However, it has been reported that the market presence of HOPCOMS has diminished over the years. The advent of retail outlet chains and reluctance of farmers to supply goods to the society has reportedly reduced the business volume from 34,000 tonnes of fruits and vegetables in 1997–98 to 20,726 tonnes during 2013–14.

Commercial Horticultural Association

The Commercial Horticultural Association (CHA) is the UK Trade Association representing and supporting companies that manufacture and supply plants, products and services to commercial horticultural growers throughout the world.

CHA was established in 1978 under the name Commercial Horticultural Exhibitors Group (CHEG) to assist exhibitors at UK Trade show to negotiate collectively with exhibition organisers. In 19?? it expanded its remit to become a trade Association and the name changed to the Commercial Horticultural Association.

The CHA was for many years administered by Brian Dunsby from offices in Harrogate, Yorkshire. When Brian retired in 1997 the administration was taken over by Groundcover Services based at Stoneleigh Park, Kenilworth, Warwichshire. Finally in 2009 the administration passed to the Federation of Garden and Leisure Manufacturers based in Brasted, Kent. Day-to-day administration remains under the care of Stuart Booker who retains an office at Stoneleigh Park.

Objectives

The Association represent the interests of exhibitors at British trade shows and selected markets overseas and to answering queries and promoting its members products and services and British horticulture as a whole throughout the world. The Association's prime functions are fact-finding, liaison and co-ordination.

Its Mission is; "to help improve the business environment by identifying the needs of companies within the horticultural supplies sector and to help them to promote their products in the UK and overseas"

The CHA publishes an annual Buyers Guide and Members' Handbook which is circulated at horticultural shows and exhibitions in the UK and overseas as well as through overseas embassies, international journals and to individual enquirers.

Urban Horticulture

Urban horticulture specifically is the study of the relationship between plants and the urban environment. It focuses on the functional use of horticulture so as to maintain and improve the surrounding urban area. With the expansion of cities and rapid urbanization, this field of study is large and complex and its study has only recently gained momentum. It has an undeniable relationship to production horticulture in that fruits, vegetables and other plants are grown for harvest, aesthetic, architectural, recreational and psychological purposes, but it extends far beyond these benefits. The value of plants in the urban environment has yet to be thoroughly researched or quantified.

Horticulture and the integration of nature into our civilization has been a major part in the establishment of our cities. When nomadic civilizations began settling down, their major trading centers were the market gardens and farms. Urban horticulture rapidly progressed with the birth of cities and the increase in experimentation and exchange of ideas. These insights led to the field being dispersed to farmers in the hinterlands. For centuries, the built environment such as homes, public buildings, etc. were integrated with cultivation in the form of gardens, farms, and grazing lands, Kitchen gardens, farms, common grazing land, etc. Therefore, horticulture was a regular part of everyday life in the city. With the Industrial Revolution and the related increasing populations rapidly changed the landscape and replaced green spaces with brick and asphalt. After the nineteenth century, Horticulture was then selectively restored in some urban spaces as a response to the unhealthy conditions of factory neighborhoods and cities began seeing the development of parks.

Post World War II trends

Early urban horticulture movements majorly served the purposes of short term welfare during recession periods, philanthropic charity to uplift "the masses" or patriotic relief. The tradition of urban horticulture mostly declined after World War II as suburbs became the focus of residential and commercial growth. Most of the economically stable population moved out of the cities into the suburbs, leaving only slums and ghettos at the city centers. However, there were a few exceptions of garden projects initiated by public housing authorities in the 1950s and 1960s for the purpose of beautification and tenant pride. But for the most part as businesses also left the metropolitan areas, it generated wastelands and areas of segregated poverty.

Inevitably the disinvestment of major city centers, specifically in America resulted in the drastic increase of vacant lots. Existing buildings became uninhabitable, houses were abandoned and even productive industrial land became vacant. Modern community gardening, urban agriculture, and food security movements were a form of response to battle the above problems at a local level. In fact other movements at that time such as the peace, environmental, women's, civil rights, and "back-to-the-city" movements of the 1960s and 1970s and the environmental justice movement of the 1980s and 1990s saw opportunity in these vacant lands as a way of reviving communities through school and community gardens, farmers' markets, and urban agriculture.

Modern community garden movement

Things have taken a turn in the twenty-first century as people are recognizing the need for local community gardens and green spaces. It is not the concept but the purposes that are new. The main goals of this movement include cleaning up neighborhoods, pushing out drug dealing that occurs at empty lots, growing and preserving food for consumption, restoring nature to industrial areas, and bringing the farming traditions to urban cities. Essentially community gardening is seen as way of creating a relationship between people and a place through social and physical engagement. Most urban gardens are created on vacant land that vary in size and are generally gardened as individual plots by community members. Such areas can support social, cultural, and artistic events and contribute to the rebuilding of local community spirit. The modern community garden movement is initiated by neighborhoods along with the support of the governments and non-profit organizations.

Some gardens are linked to public housing projects, schools through garden-based learning programs, churches and social agencies and some even employ those who are incarcerated. Community gardens which are now a large part of the urban horticulture movement are different from the earlier periods of grand park development in that the latter only served to free the people from the industrialism. In addition a community garden is more beneficial and engaging than a mere lawn or park and serves as a valuable access to nature where wilderness is unavailable. This movement helped create and sustain relationships between city dwellers and the soil and contributed to a different kind of urban environmentalism that did not have any characteristics of reform charity.

Despite that it has been 30 years since the first community gardens in the US, there is no concrete analysis of current urban gardens and their organizations. The American Community Gardening Association (ACGA) has estimations that show that municipal governments and non-profit organizations operate gardening programs in about 250 cities and towns, although the staff of this organization admits that this number could in reality be twice as large. In 1994 survey, the National Gardening Association found that 6.7 million households, that weren't involved in gardening would be interested in doing so if there was a plot nearby. A more recent survey showed that more gardens are being created in cities as opposed to being lost to economic development.

Today urban horticulture has several components that include more than just community gardens, such as market gardens, small farms and farmers' markets and is an important aspect of community development. Another result of urban horticulture is the food security movement where locally grown food is given precedence through several projects and programs, thus providing low-cost and nutritious food. Urban community gardens and the food security movement was a response to the problems of industrial agriculture and to solve its related problems of price inflation, lack of supermarkets, food scarcity, etc.

Benefits

Horticulture by itself is a practical and applied science, which means it can have a significance in our everyday lives. As community gardens cannot actually compete with market-based land uses, it is essential to find other ways to understand their various benefits such as their contribution to social, human, and financial well-being. Frederick Law Olmsted, the designer of New York City's Central Park observed that the trees, meadows, ponds and wildlife tranquilize the stresses of city life. According to various studies over the years, nature has a very positive impact over human health and even more so in an emotional and psychological sense. Trees, grass, and flower gardens, due to their presence as well as visibility, increase people's life satisfaction by reducing fatigue and irritation and restoring a sense of calm.

In fact Honeyman tested the restorative value of nature scenes in urban settings and discovered that vegetation in an urban setting produced more mental restoration as opposed to areas without vegetation. In addition, areas with only nature did not have as much of a positive psychological impact as did the combination of urban areas and nature.

One of the obvious health benefits of gardening is the increased intake of fruits and vegetables. But the act of gardening itself, is also a major health benefit. Gardening is a low-impact exercise, which when added into daily activities, can help reduce weight, lower stress, and improve overall health. A recent study showed a reduced body mass index and lower weight in community gardeners compared with their non-gardening counterparts The study showed men who gardened had a body mass index 2.36 lower and were 62% less likely to be overweight than their neighbors, while women were 46% less likely to be overweight with a body mass index 1.88 lower than their neighbors. Access to urban gardens can improve health through nutritious, edible plantings, as well by getting people outside and promoting more activity in their environments.

Gardening programs in inner-city schools have become increasingly popular as a way to teach children not only about healthy eating habits, but also to encourage students to become active learners. Besides getting students outside and moving, and encouraging an active lifestyle, children also learn leadership, teamwork, communication and collaboration skills, in addition to critical and creative thinking skills. Gardening in schools will enable children to share with their families the health and nutrition benefits of eating fresh fruits and vegetables. Because weather and soil conditions are in a state of constant change, students learn to adapt their thinking and creatively problem solve, depending on the situations that arise. Students also learn to interact and communicate with a diverse population of people, from other students to adult volunteers. These programs benefit students' health and enable them to be active contributors in the world around them.

Gardens and other green spaces also increase social activity and help in creating a sense of place, apart from their various other purposes such as enhancing the community by mediating environmental factors. There is also a huge disparity in the availability of sources that provide nutritious and affordable foods especially around urban centers which have problems of poverty, lack of public transport and abandonment by supermarkets. Therefore, inner city community gardens can be a valuable source of nutrition at an affordable cost in the most easily accessible way.

In order to understand and thereby maximize the benefits of urban horticulture, it is essential to document the effects of horticulture activities and quantify the benefits so that governments and private industries can make the appropriate changes. Horticulturists have always been involved in the botanical and physical aspects of horticulture but an involvement in its social and emotional factors would be highly beneficial to communities, cities and to the field of horticulture and its profession. Based on this, in the 1970s, the International Society for Horticultural Science recognized this need for research on the functional use of plants in an urban setting along with the need of improved communication between scientists in this field of research and people who utilize plants.

The Commission for Urban Horticulture was established in 1982 which deals with plants grown in urban areas, management techniques, the functional use of these plants as well the shortcomings of the current lack of knowledge regarding this field. The establishment of such a commission is an important indicator that this topic has reached a level of international recognition.

Economic benefits

There are many different economic benefits from gardening from saving money purchasing food and even on the utility bills. Developing countries can spend up to 60–80 percent of income on buying food alone. In Barbara Lake, Milfront Taciano and Gavin Michaels Journal of Psychology title "The Relative Influence of Psycho-Social Factors on Urban Gardening", they say that while people are saving money on buying food, having roof top gardens are also becoming popular. Having green roofs can reduce the cost of heating in the winter and help stay cool in the summer. Green roofs also can lower the cost of roof replacement. While green roofs are an addition to urban horticulture people are eating healthy while also improving the value of their property. Other benefits include increased employment from non-commercial jobs where producers include reductions on the cost of food.

Production practices

Crops are grown in flowerpots, growbags, small gardens or larger fields, using traditional or high-tech and innovative practices. Some new techniques that have been adapted to the urban situation and tackle the main city restrictions are also documented. These include horticultural production on built-up land using various types of substrates (e.g. roof top, organic production and hydroponic/aeroponic production). Because of this, it is also known as roof-top vegetable gardening/horticulture and container vegetable gardening/horticulture.

Urban and peri-urban horticulture in Africa

A report of the United Nations Food and Agriculture Organization, Growing greener cities in Africa, states that market gardening – i.e. irrigated, commercial production of fruit and vegetables in areas designated for the purpose, or in other urban open spaces – is the single most important source of locally grown, fresh produce in 10 out of 27 African countries for which data are available. Market gardening produces most of all the leafy vegetables consumed in Accra, Dakar, Bangui, Brazzaville, Ibadan, Kinshasa and Yaoundé, cities that, between them, have a total population of 22.5 million. Market gardens provide around half of the leafy vegetable supply in Addis Ababa, Bissau and Libreville.

The report says that in most of urban Africa, market gardening is an informal and often illegal activity, which has grown with little official recognition, regulation or support. Most gardeners have no formal title to their land, and many lose it overnight. Land suitable for horticulture is being taken for housing, industry and infrastructure. To maximize earnings from insecure livelihoods, many gardeners are overusing pesticide and urban waste water.

2

VEGETABLE FARMING BUSINESS

Vegetables are very important sources of vitamins, minerals, and antioxidants providing human health benefits. Vegetable farming business is a profitable business and this not only for a big farmer. It is also profitable for small and marginal farmers. A small-scale vegetable farming has the earning potential throughout the year. If you are planning for commercial vegetable production for maximum profits, you must have a proper vegetable farming business plan.

Vegetable Farming Business Plan

Growing vegetable crops is the perfect way to turn your gardening skills and knowledge into extra income. Business planning is the key to success when you're ready to invest in starting a vegetable-production business. Poor management and lack of planning are, in many cases, the main causes of business failure. Vegetable farming is a type of crop production intended mainly for human consumption of the crop's edible parts such as the shoot, leaves, fruits, and roots. According to the consuming part of the crop, vegetables are mainly divided into the following groups;

- Leafy vegetables (lettuce, cabbage, spinach)
- Fruit vegetables (pepper, cucumber, tomato)
- Root vegetables (carrot, radish, sweet potato)
- Bulb vegetables (garlic, onion, fennel)
- Flower vegetables (artichoke, cauliflower, broccoli)

Things to consider in starting a vegetable farming business

Starting a vegetable farming business.

Vegetable farming business demands proper planning, investment, adequate knowledge, and marketing. However, here we have discussed some of the main essentials;

- First of all, a solid vegetable farming business plan is very important.
- In starting, figure out how must land area you have for vegetable farming.
- According to the agro-climatic condition choose the vegetable for farming.
- You must consider the local market because vegetables are hugely perishable items.
- Also, cultivate the scope of export.
- Select the right species.
- Furthermore, you must arrange the proper irrigation for your vegetable farm.
- Plan for harvesting storage.
- Calculate the entire working capital cost.
- Finally, you must arrange the required finance.

Production factors and techniques for vegetable farming business

Vegetable farming business requires attention to all production operations, including insect, disease, and weed control and efficient marketing. The kind of vegetable grown is determined by consumer demands, which can be defined in terms of vegetable variety, size, tenderness, flavor, and type of pack. Though, effective management involves the adoption of methods resulting in a steady flow of the desired amount of produce over the whole of the natural growing season of the vegetable crop. Many vegetable plants can be grown throughout the year in some climates, while yield per acre for a given kind of vegetable varies based on the growing season and region where the crop is produced.

Climate – Climate involves the temperature level, moisture, daylight, and wind conditions of a specific region. Climatic factors strongly affect all stages and processes of vegetable plant growth

Temperature – Temperature requirements are mainly based on the minimum, optimum, and maximum temperatures during both day and night throughout plant growth.

Moisture – The amount and annual distribution of rainfall in a region, particularly during certain periods of development, affects local crops.

Daylight – Light is the source of energy for vegetable plants. The response of plants to light is mainly dependent upon light intensity, quality, and daily duration.

Site – The choice of a site involves such factors as soil and climatic regions.

Soil preparation and management – Soil preparation and management for vegetable growing involves many of the usual operations required for other crops. Good drainage is important for early vegetables because of wet soil retards development.

Propagation – Propagation of vegetable plants, involving the formation and development of new individuals in the establishment of new plantings, is accomplished by the use of either seeds or the vegetative parts of plants.

Planting – Vegetable crops are planted in the field where they are to grow to maturity. A few kinds are commonly started in a seedbed, established in the greenhouse or the open, and transplanted as seedlings.

Cultivation – Vegetable cultivation refers to stirring the soil between rows of vegetable plants.

Irrigation – Vegetable farming requires irrigation in arid and semi-arid regions, and irrigation is frequently used as insurance against drought in more humid regions.

Disease and insect control – The vegetable production of satisfactory crops requires rigorous disease- and insect-control measures. Crop yield can be lowered by disease or insect attack, and when plants are attacked at an early stage of growth the entire crop may be lost. Reduction in the quality of crops may also be caused by diseases and insects.

Harvesting – The development stage of vegetables when harvested affects the quality of the product reaching the consumer.

Marketing strategies used in a small vegetable farming business plan

Marketing strategy to the small vegetable growing farmer can be;

- Collective approaches, no individual side marketing
- Growing quality vegetables.
- Collection through cooperative or committee.
- Standardization of the product.
- Sale in the outlet by cooperative or Malls.
- Welfare strategy for farmers in profit distribution.
- Government subsidy to the collective approach.

Types of production

Vegetable production operations range from small patches of crops, producing a few vegetables for family use or marketing, to the great, highly organized and mechanized farms common in the most technologically advanced countries.

In technologically developed countries the three main types of vegetable farming are based on production of vegetables for the fresh market, for canning, freezing, dehydration, and pickling, and to obtain seeds for planting.

Production for the fresh market

This type of vegetable farming is normally divided into home gardening, market gardening, truck farming, and vegetable forcing.

Home gardening provides vegetables exclusively for family use. About one-fourth of an acre (one-tenth of a hectare) of land is required to supply a family of six. The most suitable vegetables are those producing a large yield per unit of area. Bean, cabbage, carrot, leek, lettuce, onion, parsley, pea, pepper, radish, spinach, and tomato are desirable home garden crops.

Market gardening produces assorted vegetables for a local market. The development of good roads and of motor trucks has rapidly extended available markets; the market gardener, no longer forced to confine his operations to his local market, often is able to specialize in the production of a few, rather than an assortment, of vegetables; a transformation that provides the basis for a distinction between market and truck gardening in the mid-20th century. Truck gardens produce specific vegetables in relatively large quantities for distant markets.

In the method known as forcing, vegetables are produced out of their normal season of outdoor production under forcing structures that admit light and induce favourable environmental conditions for plant growth. Greenhouses, cold frames, and hotbeds are common structures used. Hydroponics, sometimes called soilless culture, allows the grower to practice automatic watering and fertilizing, thus reducing the cost of labour. To successfully compete with other fresh market producers, greenhouse vegetable growers must either produce crops when the outdoor supply is limited or produce quality products commanding premium prices.

Production for processing

Processed vegetables include canned, frozen, dehydrated, and pickled products. The cost of production per unit area of land and per ton is usually less for processing crops than for the same crops grown for market because raw material appearance is not a major quality factor in processing. This difference allows lower land value, less hand labour, and lower handling cost. Although many kinds of vegetables can be processed, there are marked varietal differences within each species in adaptability to a given method.

Specifications for vegetables for canning and freezing usually include small size, high quality, and uniformity. For many kinds of vegetables, a series of varieties having different dates of maturity is required to ensure a constant supply of raw material, thus enabling the factory to operate with an even flow of input over a long period. Acceptable processed vegetables should have a taste, odour, and appearance comparable with the fresh product, retain nutritive values, and have good storage stability.

Vegetables raised for seed production

This type of vegetable farming requires special skills and techniques. The crop is not ready for harvest when the edible portion of the plant reaches the stage of maturity; it must be carried through further stages of growth. Production under isolated conditions ensures the purity of seed yield. Special techniques are applied during the stage of flowering and seed development and also in harvesting and threshing the seeds.

Production factors and techniques

Profitable vegetable farming requires attention to all production operations, including insect, disease, and weed control and efficient marketing. The kind of vegetable grown is mainly determined by consumer demands, which can be defined in terms of variety, size, tenderness, flavour, freshness, and type of pack. Effective management involves the adoption of techniques resulting in a steady flow of the desired amount of produce over the whole of the natural growing season of the crop. Many vegetables can be grown throughout the year in some climates, although yield per acre for a given kind of vegetable varies according to the growing season and region where the crop is produced.

Climate

Climate involves the temperature, moisture, daylight, and wind conditions of a specific region. Climatic factors strongly affect all stages and processes of plant growth.

Temperature

Temperature requirements are based on the minimum, optimum, and maximum temperatures during both day and night throughout the period of plant growth. Requirements vary according to the type and variety of the specific crop. Based on their optimum temperature ranges, vegetables may be classed as cool-season or warm-season types. Cool-season vegetables thrive in areas where the mean daily temperature does not rise above 70° F (21° C). This group includes the artichoke, beet, broccoli, brussels sprouts, cabbage, carrot, cauliflower, celery, garlic, leek, lettuce, onion, parsley, pea, potato, radish, spinach, and turnip. Warm-season vegetables, requiring mean daily temperature of 70° F or above, are intolerant of frost. These include the bean, cucumber, eggplant, lima bean, okra, muskmelon, pepper, squash, sweet corn (maize), sweet potato, tomato, and watermelon.

Premature seeding, or bolting, is an undesirable condition that is sometimes seen in fields of cabbage, celery, lettuce, onion, and spinach. The condition occurs when the plant goes into the seeding stage before the edible portion reaches a marketable size. Bolting is attributed to either extremely low or high temperature conditions in combination with inherited traits. Specific vegetable strains or varieties may exhibit significant differences in their tendency to bolt.

Young cabbage or onion plants of relatively large size may bolt upon exposure to low temperatures near 50° to 55° F (10° to 13° C). At high temperatures of 70° to 80° F (21° to 27° C) lettuce plants do not form heads and will show premature seeding. The fruit sets of tomatoes are adversely affected by relatively low and relatively high temperatures. Tomato breeders, however, have developed several new varieties, some setting fruits at a temperature as low as 40° F (4° C) and others at a temperature as high as 90° F (32° C).

Moisture

The amount and annual distribution of rainfall in a region, especially during certain periods of development, affects local crops. Irrigation may be required to compensate for insufficient rainfall. For optimum growth and development, plants require soil that supplies water as well as nutrients dissolved in water. Root growth determines the extent of a plant's ability to absorb water and nutrients, and in dry soil root growth is greatly retarded. Extremely wet soil also retards root growth by restricting aeration. Atmospheric humidity, the moisture content of the air, also contributes moisture. Certain seacoast areas characterized by high humidity are considered especially adapted to the production of such crops as the artichoke and lima bean. High humidity, however, also creates conditions favourable for the development of certain plant diseases.

Daylight

Light is the source of energy for plants. The response of plants to light is dependent upon light intensity, quality, and daily duration, or photoperiod. The seasonal variation in day length affects the growth and flowering of certain vegetable crops. Continuation of vegetative growth, rather than early flower formation, is desirable in such crops as spinach and lettuce. When planted very late in the spring, these crops tend to produce flowers and seeds during the long days of summer before they attain sufficient vegetative growth to produce maximum yields. The minimum photoperiod required for formation of bulbs in garlic and onion plants differs among varieties, and local day length is a determining factor in the selection of varieties.

Each of the climatic factors affects plant growth, and can be a limiting factor in plant development. Unless each factor is of optimum quantity or quality, plants do not achieve maximum growth. In addition to the importance of individual climatic factors, the interrelationship of all environmental factors affects growth.

Certain combinations may exert specific effects. Lettuce usually forms a seedstalk during the long days of summer, but the appearance of flowers may be delayed, or even prevented, by relatively low temperature. An unfavourable temperature combined with unfavourable moisture conditions may cause the dropping of the buds, flowers, and small fruits of the pepper, reducing the crop yield. Desirable areas for muskmelon production are characterized by low humidity combined with high temperature. In the production of seeds of many kinds of vegetables, absence of rain, or relatively light rainfall, and low humidity during ripening, harvesting, and curing of the seeds are very important.

Site

The choice of a site involves such factors as soil and climatic region. In addition, with the continued trend toward specialization and mechanization, relatively large areas are required for commercial production, and adequate water supply and transportation facilities are essential. Topography—that is, the surface of the soil and its relation to other areas—influences efficiency of operation. In modern mechanized farming, large, relatively level fields allow for lower operating costs. Power equipment may be used to modify topography, but the cost of such land renovation may be prohibitive. The amount of slope influences the type of culture possible. Fields with a moderate slope should be contoured, a process that may involve added expense for the building of terraces and diversion ditches. The direction of a slope may influence the maturation time of a crop or may result in drought, winter injury, or wind damage. A level site is generally most desirable, although a slight slope may assist drainage. Exposed sites are not suitable for vegetable farming because of the risk of damage to plants by strong winds.

The soil stores mineral nutrients and water used by plants, as well as housing their roots. There are two general kinds of soils—mineral and the organic type called muck or peat. Mineral soils include sandy, loamy, and clayey types. Sandy and loamy soils are usually preferred for vegetable production. Soil reaction and degree of fertility can be determined by chemical analysis. The reaction of the soil determines to a great extent the availability of most plant nutrients. The degree of acid, alkaline, or neutral reaction of a soil is expressed as the pH, with a pH of 7 being neutral, points below 7 being acid, and those above 7 being alkaline. The optimum pH range for plant growth varies from one crop to another. A soil can be made more acid, or less alkaline, by applying an acid-producing chemical fertilizer such as ammonium sulfate.

The inherent fertility of soils affects production quantity, and a sound fertility program is required to maintain productivity. The ability of a soil to support plant life and produce abundant harvests is dependent on the immediately available nutrients in the soil and on the rate of release of additional nutrients that are present but not available to plants. The rate of release of these additional nutrients is affected by such factors as microbial action, soil temperature, soil moisture, and aeration. Depletion of soil fertility may occur as a result of crop removal, erosion, leaching, and volatilization, or evaporation, of nutrients.

Soil preparation and management

Soil preparation for vegetable growing involves many of the usual operations required for other crops. Good drainage is especially important for early vegetables because wet soil retards development. Sands are valuable in growing early vegetables because they are more readily drained than the heavier soils. Soil drainage accomplished by means of ditches or tiles is more desirable than the drainage obtained by planting crops on ridges because the former not only removes the excess water but also allows air to enter the soil. Air is essential to the growth of crop plants and to certain beneficial soil organisms making nutrients available to the plants.

When crops are grown in succession, soil rarely needs to be plowed more than once each year. Plowing incorporates sod, green-manure crops, and crop residues in the soil; destroys weeds and insects; and improves soil texture and aeration. Soils for vegetables should be fairly deep. A depth of six to eight inches (15 to 20 centimetres) is sufficient in most soils.

Soil management involves the exercise of human judgment in the application of available knowledge of crop production, soil conservation, and economics. Management should be directed toward producing the desired crops with a minimum of labour. Control of soil erosion, maintenance of soil organic matter, the adoption of crop rotation, and clean culture are considered important soil-management practices.

Soil erosion, caused by water and wind, is a problem in many vegetable-growing regions because the topsoil is usually the richest in fertility and organic matter. Soil erosion by water can be controlled by various methods. Terracing divides the land into separate drainage areas, with each area having its own waterway above the terrace. The terrace holds the water on the land, allowing it to soak into the soil and reducing or preventing gullying. In the contouring system, crops are planted in rows at the same level across the field. Cultivation proceeds along the rows rather than up and down the hill. Strip cropping consists of growing crops in narrow strips across a slope, usually on the contour. Soil erosion by wind can be controlled by the use of windbreaks of various kinds, by keeping the soil well supplied with humus, and by growing cover crops to hold the soil when the land is not occupied by other crops.

Maintenance of the organic-matter content of the soil is essential. Organic matter is a source of plant nutrients and is valuable for its effect on certain properties of the soil. Loss of organic matter is the result of the action of micro-organisms that gradually decompose it to carbon dioxide. The addition of manures and the growing of soil-improving crops are efficient means of supplying soil organic matter. Soil-improving crops are grown solely for the purpose of preparing the soil for the growth of succeeding crops. Green-manure crops, grown especially for soil improvement, are turned under while still green and usually are grown during the same season of the year as the vegetable crops. Cover crops, raised for both soil protection and improvement, are only grown during seasons when vegetable crops do not occupy the land. When a soil-improving crop is turned under, the various nutrients that have contributed to the growth of the crop are returned to the soil, adding a quantity of organic matter. Both legumes, those plants such as peas and beans having fruits and seeds formed in pods, and nonlegumes are effective soil-improving crops. The legumes, however, are more valuable, because they contribute nitrogen as well as humus. The rate of decomposition of plant material depends on the kind of crop, its stage of growth, and soil temperature and moisture. The more succulent the material is at the time it is turned under, the more quickly it decomposes. Because dry material decomposes more slowly than green material, it is desirable to turn under soil-improving crops before they are mature, unless considerable time is to elapse between the plowing and the planting of the succeeding crop. Plant material decomposes most rapidly when the soil is warm and well supplied with moisture. If soil is dry when a soil-improving crop is turned under, little or no decomposition will occur until rain or irrigation supplies the necessary moisture.

The chief benefits derived from crop rotation are the control of disease and insects and the better use of the resources of the soil. Rotation is a systematic arrangement for the growing of different crops in a more or less regular sequence on the same land. It differs from succession cropping in that rotation cropping covers a period of two, three, or more years, while in succession cropping two or more crops are grown on the same land in one year. In many regions vegetable crops are grown in rotation with other farm crops. Most vegetables grown as annual crops fit into a four-or five-year rotation plan. The system of intercropping, or companion cropping, involves the growing of two or more kinds of vegetables on the same land in the same growing season. One of the vegetables must be a small-growing and quick-maturing crop; the other must be larger and late maturing.

In the practice of clean culture, commonly followed in vegetable growing, the soil is kept free of all competing plants through frequent cultivation and the use of protective coverings, or mulches, and weed killers. In a clean vegetable field the possibility of attack by insects and disease-incitant organisms, for which plant weeds serve as hosts, is reduced.

Propagation

Propagation of crop plants, involving the formation and development of new individuals in the establishment of new plantings, is usually accomplished by the use of either seeds or the vegetative parts of plants. The first type, known as sexual propagation, is used for asparagus, bean, broccoli, cabbage, carrot, cauliflower, celery, cucumber, eggplant, leek, lettuce, lima bean, okra, onion, muskmelon, parsley, pea, pepper, pumpkin, radish, spinach, sweet corn (maize), squash, tomato, turnip, and watermelon. The second type, asexual propagation, is used for the artichoke, garlic, girasole, potato, rhubarb, and sweet potato.

Although seed cost is a small portion of the total cost of crop production, seed quality strongly affects crop success or failure. Good seed should be accurately labelled, clean, graded to size, viable, and free of diseases and insects. The reliability of the seed house is an important factor in obtaining good-quality seed. Viability, or ability to grow, and longevity, the period of viability, are characteristics of seeds of any vegetable kind. In cool, dry storage conditions, those vegetable seeds having comparatively short longevity of one to two years are okra, onion, parsley, and sweet corn. Seeds having three-year longevity are those of the asparagus, bean, carrot, leek, and pea; four-year longevity is characteristic of the beet, chard, pepper, pumpkin, and tomato seeds; longevity of five years characterizes the seeds of broccoli, cabbage, cauliflower, celery, cucumber, eggplant, lettuce, muskmelon, radish, spinach, squash, turnip, and watermelon. The dry seeds of all vegetables, when packed under vacuum in hermetically sealed cans, should remain viable for a longer period than seeds stored under less protective conditions.

Crops grown from hybrid seeds (the offspring of two or more selected parental varieties and known as F1) yield vegetables of high quantity and quality. The hybrid-seed industry is based on the production of new seed each year from the controlled pollination of selected parents found to produce the desired combination of characters in the progeny. In the early 1980s the number of F1 hybrids was increasing in Japan, the United States, and other technically advanced countries. The number of F1 hybrids varied with the kind of vegetable, but none had yet been introduced for the bean, celery, lettuce, okra, parsley, or pea.

Planting

Most vegetable crops are planted in the field where they are to grow to maturity. A few kinds are commonly started in a seedbed, established in the greenhouse or in the open, and transplanted as seedlings. Asparagus seeds are planted in a seedbed to produce crowns used for field setting. Some vegetables can be either directly seeded in the field or grown from transplants. These include broccoli, cabbage, cauliflower, celery, eggplant, leek, lettuce, onion, pepper, and tomato. The time and method of planting seeds and plants of a particular vegetable influence the success or failure of the crop. Important factors include the depth of planting, the rate of planting, and the spacing both between rows and between plants within a row.

Factors to be considered in determining the time of planting include soil and weather conditions, kind of crop, and desired harvest time. When more than one planting of a crop is made, the second and later plantings should be timed to provide a continuous harvest for the period desired. The soil temperature required for germination of the planted seed varies markedly with the various kinds of vegetables. Vegetables that will not germinate at a temperature below 60° F (16° C) include the bean, cucumber, eggplant, lima bean, muskmelon, okra, pepper, pumpkin, squash, and watermelon. Temperatures higher than 90° F (32° C) are not favourable for the germination of seeds of celery, lettuce, lima bean, parsley, pea, and spinach.

The quantity of seeds planted, or rate of planting, is mainly determined by the characteristics of the vegetable plant. The size of seeds affects the number of plants raised in a given area. Watermelon varieties, for example, differ in seed size expressed as weight. The Sugar Baby variety has an average weight of 1.4 ounces (41 grams) for 1,000 seeds;

those of Blackstone variety average 4.4 ounces (125 grams). If the two are grown on two separate plots of the same area and 4.4 ounces of seeds of each cultivar are planted, the result would be three times as many of the Sugar Baby plants as the Blackstone type. Seed size and plant-growth pattern of a vegetable are major factors that govern the number of plants raised in a given area. The trend in the early 1980s was to increase plant population for many crops to achieve the greatest yield possible without impairing quality. As plant population increases per unit area, a point is reached at which each plant begins to compete for certain essential growth factors—e.g., nutrients, moisture, and light. When the population is below the level in which competition between plants occurs, increased population will have no effect on individual plant performance, and the yield per unit area will increase in direct proportion to the increment of population. When competition for essential growth factors occurs, however, yield per plant decreases.

Early harvest and economical use of space are the principal objectives of growing vegetable crops from transplants produced in a greenhouse or outdoor seedbed. It is easier to care for young plants of the cabbage, cauliflower, celery, onion, and tomato in small seedbeds than to sow the seeds in the place where the crop is to grow and mature. Land is free longer for another crop, and weeds, insects, diseases, and irrigation are more readily and economically controlled. The production of transplants is often a specialty of growers who sell their produce to other vegetable growers. The seeds may be planted at a rate three to six times that commonly used for a direct-seeded field. The young plants are removed for use as transplants when they reach the desired size and age, approximately 40 to 60 days after seeding.

Care of crops during growth

Practices required for a vegetable crop growing in the field include cultivation; irrigation; application of fertilizers; control of weeds, diseases, and insects; protection against frost; and the application of growth regulators if necessary.

Cultivation

Cultivation refers to stirring the soil between rows of vegetable plants. Because weed control is the most important function of cultivation, this work should be performed at the most favourable time for weed killing, when the weeds are breaking through the soil surface. When the plants are grown on ridges, it is necessary to cover the basal plant portion with soil in the case of such vegetables as asparagus, carrot, garlic, leek, onion, potato, sweet corn, and sweet potato.

Irrigation

Vegetable production requires irrigation in arid and semi-arid regions, and irrigation is frequently used as insurance against drought in more humid regions. In areas having intermittent rain for five or six months, with little or none during the remainder of the year, irrigation is essential throughout the dry season and may also be needed between rainfalls in the rainy season. The two types of land irrigation generally suited to vegetables are surface irrigation and sprinkler irrigation. A level site is required for surface irrigation, in which the water is conveyed directly over the field in open ditches at a slow, nonerosive velocity. Where water is scarce, pipelines may be used, eliminating losses caused by seepage and evaporation. The distribution of water is accomplished by various control structures, and the furrow method of surface irrigation is frequently employed because most vegetable crops are grown in rows. Sprinkler irrigation conveys water through pipes for distribution under pressure as simulated rain.

Irrigation requirements are determined by both soil and plant factors. Soil factors include texture, structure, water-holding capacity, fertility, salinity, aeration, drainage, and temperature. Plant factors include type of vegetable, density and depth of the root system, stage of growth, drought tolerance, and plant population.

Fertilizer application

Soil fertility is the capacity of the soil to supply the nutrients necessary for good crop production, and fertilizing is the addition of nutrients to the soil. Chemical fertilizers may be used to supply the needed nitrogen, phosphorus, and potassium. Chemical tests of soil, plant, or both are used to determine fertilizer needs, and the rate of application is usually based on the fertility of the soil, the cropping system employed, the kind of vegetable to be grown, and the financial return that might be expected from the crop. Methods of fertilizer application include scattering and mixing with the soil before planting; application with a drill below the surface of the soil at the time of planting; row application before or at planting time; and row application during plant growth, also called side-dressing. Plowed down broadcast fertilizers have recently been used in combination with high analysis liquid fertilizers applied at planting or as a side-dressed band. Mechanical planting devices may employ fertilizer attachments to plant the fertilizer in the form of bands near the seed. For most vegetables, the bands are placed from two to three inches (five to 7.5 centimetres) from the seed, either at the same depth or slightly below the seed.

Weed control

Weeds (plants growing where they are not wanted) reduce crop yield, increase production cost, and may harbour insects and diseases that attack crop plants. Methods employed to control weeds include hand weeding, mechanical cultivation, application of chemicals acting as herbicides, and a combination of mechanical and chemical means. Herbicides, selective chemical weed killers, are absorbed by the plant and induce a toxic reaction. The amount and type of herbicide that can be safely used to protect vegetable crops depends on the tolerance of the specific crops to the chemical. Most herbicides are applied as a spray, and the appropriate time for application is determined by the composition of the herbicide and the kind of vegetable crop to be treated. Preplanting treatments are applied before the crop is planted; preemergence treatments are applied after the crop is planted but before its seedlings emerge from the soil; and postemergence treatments are applied to the growing crop at a definite stage of growth.

Disease and insect control

The production of satisfactory crops requires rigorous disease- and insect-control measures. Crop yield may be lowered by disease or insect attack, and when plants are attacked at an early stage of growth the entire crop may be lost. Reduction in the quality of vegetable crops may also be caused by diseases and insects. Grades and standards for market vegetables usually specify strict limits on the amount of disease and insect injury that may be present on vegetables in a designated grade. Vegetables remain vulnerable to insect and disease damage after harvesting, during the marketing and handling processes. When a particular plant pest is identified, the grower can select and apply appropriate control measures. Application of insect control at the times specific insects usually appear or when the first insects are noticed is usually most effective. Effective disease control usually requires preventive procedures.

Diseases are incited by such living organisms as bacteria, fungi, and viruses. Harmful material enters the plant, develops during an incubation period, and finally causes infection, the reaction of the plant to the pathogen, or disease-producing organism. Control is possible during the inoculation and incubation phases, but when the plant reaches the infection stage it is already damaged. Typical plant diseases include mildew, leaf spots, rust, and wilt. Chemical fungicides may be used to control disease, but the use of disease-resistant plant varieties is the most effective means of control.

Vegetable breeders have developed plant varieties resistant to one or more diseases; such varieties are available for the bean, cabbage, cucumber, lettuce, muskmelon, onion, pea, pepper, potato, spinach, tomato, and watermelon.

Insects are usually controlled by the use of chemical insecticides that kill through toxic action. Many insecticides are toxic to harmful insects but do not affect bees, which are valuable for their role in pollination.

Frost protection

Frost protection may be accomplished by increasing the amount of heat radiated from the soil when frost is likely to occur. Irrigation on the day before a predicted frost provides additional moisture in the soil to increase the amount of heat given off as infrared rays. This extra heat protects the plants from frost injury. A continuous supply of water provided by sprinkler irrigation may also protect plants from frost. As the water freezes on the plant leaves, it loses heat that is absorbed by the plant leaves, maintaining leaf temperature at 32° F (0° C). Because of the sugars and other substances in plant cells, the freezing point of cell sap is somewhat lower than 32° F.

Growth regulators

It is sometimes desirable to retard or accelerate maturity in vegetable crops. A chemical compound may be applied to prevent sprouting in onion crops. It is applied in the field sufficiently early for absorption by the still-green foliage but late enough to avoid suppressing the bulb yield. Another substance may be used to end the dormancy, or rest period, of newly harvested potato tubers intended for planting. The treated seed potatoes have uniform sprout emergence. The same substance is applied to celery from two to three weeks before harvest to elongate the stalks and increase the yield and is also used to accelerate maturity in artichokes. A chemical compound, applied when adverse weather conditions prevail during the period of fruit setting, has been used to encourage fruit set.

Harvesting

The stage of development of vegetables when harvested affects the quality of the product reaching the consumer. In some vegetables, such as the bean and pea, optimum quality is reached well in advance of full maturity and then deteriorates, although yield continues to increase. Factors determining the harvest date include the genetic constitution of the vegetable variety, the planting date, and environmental conditions during the growing season. Successive harvest dates may be obtained either by planting varieties having different maturity dates or by changing the sequence of planting dates of one particular variety. The successive method is applicable to such crops as broccoli, cabbage, cauliflower, muskmelon, onion, pea, sweet corn (maize), tomato, and watermelon. Certain varieties of the carrot, celery, cucumber, lettuce, parsley, radish, spinach, or summer squash can be sown in succession throughout most of the year in some climates, thus prolonging the harvest period.

Hand harvesting is employed along with various mechanical aids for broccoli, cabbage, cauliflower, muskmelon, and pepper crops. Many vegetables grown for processing and some vegetables destined for the fresh market are mechanically harvested. Harvesting operations may be performed by a single machine in a single step for such vegetable crops as the bean, beet, carrot, lima bean, onion, pea, potato, radish, spinach, sweet corn, sweet potato, and tomato. Designers of harvesting machinery have been working to develop a multiple-picking harvester capable of adjustment for use with more than one crop. Vegetable breeders have been able to produce vegetables with characteristics suitable for machine harvesting, including compact plant growth, uniform development, and concentrated maturity.

Storage

Fresh vegetables are living organisms, and there is a continuation of life processes in the vegetable after harvest. Changes that occur in the harvested, nonprocessed vegetable include water loss, conversion of starches to sugars, conversion of sugars to starches, flavour changes, colour changes, toughening, vitamin gain or loss, sprouting, rooting, softening, and decay.

Some changes result in quality deterioration; others improve quality in those vegetables that complete ripening after harvest. Postharvest changes are influenced by such factors as kind of crop, air temperature and circulation, oxygen and carbon dioxide contents and relative humidity of the atmosphere, and disease-incitant organisms. To maintain the fresh vegetable in the living state, it is usually necessary to slow the life processes, though avoiding death of the tissues, which produces gross deterioration and drastic differences in flavour, texture, and appearance.

Storage of vegetables contributes to price stabilization by carrying over produce from periods of high production to periods of low production. It also extends the period of consumption of many kinds of vegetables. Storage conditions can contribute to the preservation of the natural living state of the edible portion and to the prevention of deterioration through control of temperature, relative humidity, and the quality of the produce to be stored. Vegetables for storage must be free from mechanical, insect, and disease injury and should be at the proper stage of maturity.

Common (unrefrigerated) storage and cold (refrigerated) storage are the methods generally employed for vegetables. Common storage, lacking precise control of temperature and humidity, includes the use of insulated storage houses, outdoor cellars, or mounds. Cold storage allows precise regulation of temperature and humidity and maintenance of constant conditions by use of a refrigeration and ventilation system. Temporary storage, suitable only for very brief storage periods, is frequently practiced in the shipping season when large lots are accumulated for carload or truck quantities. The refrigerator car or truck is a means of temporary storage used to protect produce while it is in transit. Short-term storage may last for four or six weeks. Economic factors, such as the probability that prices will increase later in the season, encourage long-term storage of such perishable vegetables as the onion, potato, and sweet potato.

Premarketing operations and selling

Premarketing operations include washing, trimming, waxing, precooling, grading, prepackaging, and packaging. Vegetables often require washing after harvest to remove any adhering soil particles. Such vegetables as the beet, carrot, celery, lettuce, radish, spinach, and turnip are trimmed before washing to remove discoloured leaves or to cut back the green tops. Waxing of the cucumber, muskmelon, pepper, potato, sweet potato, and tomato gives the product a bright appearance and controls shrivelling through reduction of moisture loss.

Precooling

Precooling, the rapid removal of heat from freshly harvested vegetables, allows the grower to harvest produce at optimum maturity with greater assurance that it will reach the consumer at maximum quality. Precooling benefits the vegetable by slowing the natural deterioration that starts shortly after harvest, slowing the growth of decay organisms and reducing wilt by retarding water loss. The major precooling methods include hydrocooling, contact icing, vacuum cooling, and air cooling. In hydrocooling the vegetable is cooled by direct contact with cold water flowing through the packed containers and absorbing heat directly from the produce. In contact icing crushed ice is placed in the package or spread over a stack of packages to precool the contents. The vacuum cooling process produces rapid evaporation of a small quantity of water, lowering the temperature of the crop to the desired level. Air cooling involves the exposure of vegetables to cold air; the air must be as cold as possible for rapid cooling but not low enough to freeze the produce exposed to the direct air blast.

The preferred method of precooling varies according to the physical characteristics of the vegetable. Hydrocooling is recommended for the asparagus, beet, broccoli, carrot, cauliflower, celery, muskmelon, pea, radish, summer squash, and sweet corn (maize); cabbage, lettuce, and spinach are suited to vacuum cooling; air cooling is preferred for bean, cucumber, eggplant, pepper, and tomato. After the produce is precooled, it is desirable to maintain low temperature by shipping in refrigerator cars or trucks, by storing in cold-storage rooms, and by refrigeration in retail stores.

Grading

Uniformity in size, shape, colour, and ripeness is of great importance in marketing any vegetable product, and can be secured through grading. The establishment of standard grades furnishes a basis of trade. Grade standards are based mainly on general appearance, size, trueness to type, and freedom from blemishes and defects.

Packaging

Prepackaging, or consumer packaging, has become a highly organized practice, often employing elaborate equipment. The product is placed in bags made of transparent film, trays or cartons overwrapped with transparent film, or mesh or paper bags. The packaging of produce in consumer packages lends itself to self-service in retail stores. The production region is often the most satisfactory location for prepackaging, especially when a packaging centre serves a large vegetable-growing area.

Master containers for consumer packages are commonly made of paperboard. Cartons, bags, baskets, boxes, crates, and hampers of various kinds and sizes are all used in packaging vegetables for marketing. The type of container is selected to fit the kind of vegetable; it furnishes a convenient means for transport, loading, and stacking, with security and economy of space. Uniform product throughout the package is an important consideration in packing vegetables.

Selling

Producers sell vegetables through various retail and wholesale practices. Retail sales are made directly to the consumer, often through roadside stands. Many growers sell most of their produce at wholesale to retail stores, to various types of buyers on local markets in nearby cities, or in regional markets. Growers located long distances from markets sell largely to wholesale dealers or jobbers.

Some growers have contracts with processors. Wholesale marketing arrangements are also made through auction markets in the producing regions and through cooperative organizations of producers.

Importance of Vegetable Production

Vegetables are vital to the general good health of human beings, and providing necessary vitamins and minerals, and reducing risk from dangerous diseases and other medical conditions. First, of course, you would need a piece of land to start vegetable farming and try at least an acre for commercial vegetable growing. Then you would require equipment, which you can buy, lease or borrow, such as a tractor, tiller, plow, disc, cultivator, and planter. Lower your production cost as much as possible by spending on equipment only when required. Unnecessary expenses on equipment can eat away potential profits.

Vegetable production provides a promising economic opportunity for reducing rural poverty and unemployment in developing countries and is the main component of farm diversification strategies. Vegetables are mankind's most affordable source of vitamins and minerals required for good health.

Importance of vegetable production is:

- Importance in human nutrition
- Vegetables are a very important source of farm income
- Vegetables have aesthetic value
- Vegetable production for medicinal purpose
- Roll of vegetables in the national economy
- Flexibility in plant production program-unlike the fruits with vegetables the production program can be adjusted and changed for better profits according to needs. With fruits, it is a difficult time taking and expensive to change the production program if it turns out to be unprofitable.

Factors that determine successful vegetable production

Whether the growth of vegetables is intended for fresh consumption, processing, and seed production, it can be a profitable vegetable business. However, there are a few factors that can influence the profitability of vegetable production from its early beginnings;

- **Seed quality;** the sowing of quality, clean, labeled, graded to size, viable, and healthy seed can make all the difference between success and failure in vegetable farming.

- **Optimal time of sowing and planting;** depends on the climate and environmental conditions of the specific area, as well as requirements of each crop.
- **Method of planting;** the secret to successful vegetable farming lies in the managing of optimal plant requirements, by combining the production of transplants in the greenhouses with planting in the field.
- **Finally,** considering effective farm management is the first step in creating profitable vegetable production. In essence, farming of these colorful plants can be a profitable business.
- **Some plants** have high labor requirements to grow. Before selecting a vegetable to raise, know first the extent at which some plants need tending. Then, determine whether you have the time to invest to grow and market it. For example, if you expect to be unable to get your products sold immediately, avoid easily perishable crops such as asparagus, sweet corn, peas and grow potatoes and onions instead.
- **Some plants** are difficult to grow and need special attention from the farmer for optimum results. Your choice of the crop must consider whether you have the knowledge and experience in growing such crops and whether you are willing to learn from available resources. Also, some plants would need special equipment. Select those you won't need to buy the equipment to grow.

Production techniques of quality vegetables

The quality of vegetables mainly depends on the horticultural production systems, environmental factors, and management practices used. Climatic conditions such as temperature and light intensity have a strong influence on the nutritional quality of vegetables. Hydroponic cultivation technique ensures the production of quality vegetables, and in this culture system, both plant nutrition and environmental conditions are artificially managed according to the plant need. Growing quality vegetables is easier and safer in hydroponic compared to conventional soil culture. The advantages of this system are that plant roots are visible and the root zone environment can be easily monitored. In this system of cultivation, the yield of the vegetable crop can be maximized through the efficient use of all resources, and it is believed to be the intensive form of agricultural enterprises for commercial production of greenhouse vegetable plants.

Soilless culture of vegetables uses inert organic or inorganic substrate through the hydroponic nutrient application. This culture has been reported to practice in the greenhouse as an alternative to conventional filed cultivation of many high-value vegetable crops. Under these protected cultivation systems, weather factors, the amount and composition of nutrient solution, and the growing medium can be managed successfully. Therefore, the quality of vegetable crops grown through soilless culture improves significantly compared to conventional soil culture. Many researchers found better taste, uniformity, color, texture, and higher nutritional value in fruits grown in soilless culture than in soil cultivation methods.

Production plan of a vegetable farming business

Once you have a clear idea of what you want your vegetable farm business to look like, what you want to produce, and where you will sell your product, you need to establish a production plan. Some factors to consider are listed below:

Capital needs – Identify the investment and cash operating needs and how much you will need to borrow.

Infrastructure and equipment – Identify what equipment you need for the vegetable crops you will produce. Also, depending on the packaging and also handling requirements identify what type of infrastructure will be needed.

Management – Identify the production, management, and marketing skills essential to make your enterprise successful. If you do not have those skills, identify ways to acquire them, which can include hiring additional labor.

Planting and harvesting schedule – Plan the best timing for planting and harvesting your vegetable crops, based on plant varieties and availability of labor. Remember to plan planting dates based on your harvest schedule (e.g., customer demand).

Post-harvest and sanitation – Post-harvesting needs (sanitation, handling, and cooling) are very important aspects that need careful thought. Cooling is essential to delay produce spoilage and keep it fresh. When the product is not sold and delivered immediately after harvest, a cold storage option can be needed.

Enterprise analysis – Keep good plant production and financial records to help you make good decisions in the future. Use records to identify problems that need to be solved and to identify what practices and crops are profitable for your business.

Some of the important high yield vegetable crops

List of high yield vegetable crops can be given below:

Cucumbers – In an acre area, around 12000 cucumber plants are planted (3 plants per square meter) and each plant yields an average of about 5 to 7 kg per cycle. This will yield about 8,400 to 10,500 plants per acre.

Squash – In general, each squash plant produces about 5 to 25 pounds of yellow squash during the growing season. A 10-foot row of yellow squash averages about 20 to 80 pounds of squash.

Beans – The average yield is about 100 to 120 quintals of green pods per hectare can be expected.

Tomatoes – The average tomato crop yield per acre in India is 10 tonnes although the yield varies from 15 to 20 tonnes per acre in case of irrigated crops.

Peanuts – Grown mainly through age-old farming techniques, peanut yield in India is about 700 to 900 kg per hectares.

Potatoes – During the first year of cultivating potatoes, a good yield can be about 10 tons per acre. Experienced farmers after years of practice can achieve yields 16 to 28 tons per acre.

Peppers – The yield per acre of pepper is about 0.39 tonnes per hectare. This indicates a plant population of 10,250 plants per acre, thus the average yield per plant is 3.6 pounds.

Beetroot – The beetroot crop yields about 20–25 tonnes/hectare in 120 days.

Radishes – It yields about 200 to 250 quintals fresh radish per hectare.

Lettuce – The average yield of lettuce is about 80 to 120 quintals per hectare.

How to Make Money from a Horticulture Farming

Vegetable farming is the growing of several vegetables for human consumption. Vegetable farming is a type of crop production intended mainly for human consumption of the crop's edible parts such as the shoot, leaves, fruits, and roots. Growing of vegetables starts with mainly sowing or planting. Vegetable sowing means putting a seed directly into the prepared soil and vegetable planting includes the practice of putting already grown seedlings into the soil. Seedlings could be grown in the field or in greenhouses. Noe, let's get into details of vegetable farming business to make good money.

Ideas to make money from a vegetable farming business

Growing vegetable plants is something you can do almost anywhere. You don't have to live on a farm to enjoy the benefits of making money from growing a vegetable garden. You can use your space to grow fresh and healthy vegetable plants which can earn you an exciting amount of profit. There are multiple different ways that you can make money from growing vegetables. For instance, you could have a small plot of land and want to sell products directly to your customers through roadside stands, farmer's markets, and community-supported agriculture.

Factors that determine successful vegetable production

Whether the growth of vegetable plants is intended for fresh consumption, processing, or seed production, it can be a profitable business. However, there are some factors that may influence the profitability of vegetable production;

Seed quality – sowing of seed quality, clean, labeled, graded to size, viable, and healthy seed can make all the difference between success and failure in vegetable farming

Optimal time of sowing/planting – it depends on the climate and environmental conditions of the specific area, as well as the requirements of each crop

Method of planting – the secret to successful vegetable farming lies in the managing of optimal plant requirements, by combining the production of transplants in the greenhouses with planting in the field

Growing vegetable plants for profit is a great way to turn your gardening skills into serious cash. While most of us immediately think of tomatoes or salad greens, the most profitable vegetable plants are specialty crops that are not always found in a home vegetable garden. Many specialty crops can bring as much as $90,000 per acre and are easy to grow.

Numerous ways to improve the profit or make money from vegetable farming

There are numerous ways to improve the profit from vegetable farming but without any specific condition of the field, weather, market, and other factors not being mentioned, it's going to be difficult to be specific about it.

Crop – Selecting a proper vegetable crop that has the highest potential to grow and yield better based on climatic conditions is important to get good profit.

Structure – If the weather permits growing vegetables under a Polyhouse would result in a high yield with a controlled environment.

Market – Based on the market demand or availability of export facility the vegetables can be exported or sold to buyers who export.

Organic – Growing vegetables organically would increase the profit but at the same time, the yield won't be the same as the traditional way. It has a specific niche market and it can be exploited to see the outcome based on who and whom you target.

Technology – Precision technology is best to give the plants what it needs at the right time and right amount with minimal waste and its energy efficiency too. This would indirectly help in reducing the input cost and saving money and giving a good profit.

Direct selling – Cutting the involvement of middleman could greatly increase your profit. There are lots of online apps such as Aibono or Farmerfriend which connects farmers to buyers directly. You can also visit hotels or supermarkets who would buy your product directly and this gives the farmer an advantage to ask for the right price.

Time – Time is important in vegetable production because you can see most of the vegetable rates are higher before and after the main season. If you can grow such types of vegetables which give yield 15 or 30 days early or later with time management and variety then you can earn some extra amount of money from the market. But you keep in mind whenever you grow an unseasonal crop the yield will affect but you can get extra price also.

What are the expenses of a vegetable garden?

To calculate the true cost to start a vegetable garden and maintain it throughout the year, add together the following factors;

- Cost of plants or seeds
- Cost to provide nutrient-rich soil (dirt, fertilizer, and worms)
- Cost to protect and structure plants like cages, coverings, and fences
- Cost to water plants
- Cost of tools and accessories (tiller, gloves, and spade)

How much does it cost to start a vegetable farming?

Costs vary for different types of vegetable plants. All of the expenses of growing and selling vegetables fall into mainly two categories;

- Controllable and
- Fixed

Fixed expenses contain costs that do not change from year to year, while controllable expenses change based on how many acres of vegetables you plan to grow and sell. The fixed costs include buying or leasing land to grow vegetables. If you take out a loan to fund the business, include that in the fixed cost estimate. Hiring workers to take care of the field means adding in the costs of worker's compensation and also liability insurance as a fixed cost as well. Everything else in the cost estimate, including workers' hourly wages, is a controllable expense.

First, calculate the costs of the seeds or transplants of the vegetables you want to plant. Also include the labor involved in laying down mulch between the rows, planting the field, fertilizing the vegetable plants, weeding, and pest control. If your fields do not receive enough water during the growing season, account for the cost of irrigation, including the equipment you install and the amount of water the vegetable plants need to thrive and grow. Also, costs involved with planting and care include items such as string or twine on which to drape vines and any fencing necessary to keep domestic or wild animals from grazing on your vegetable crops.

Harvesting costs involve heavy labor to pick, sort, and pack your crops and some vegetables require multiple harvesting sessions over a few weeks. Other harvesting expenses contain containers to hold the produce and fuel to move the containers of produce out of the field and into buyers' hands, into storage, or to a farmers' market or produce stand.

For example, the cost of cultivation of onion per acre

If you are ready to earn the money through Onions farming, below are the cost details;

1. Cost of seeds -5,200/-
2. Labor on land preparation – 1,200/-
3. Machine Labor – 1500/-
4. Labor cost for planting – 1750/-
5. Weeding – 700/-
6. Labor for irrigation – 800/-
7. Fertilizers and Farm Yard Manure -800/-
8. Plant Protecting Chemicals- 3,600/-
9. Harvesting Charges – 1,100/-
10. Labor on Shed curing – 5,000/-

Sub Total (A) Variable Cost – 21,650/-

11. The rental value of Land – 3,300/-
12. The interest of input charges – 900/-

Sub Total (B) Fixed charges – 4,200/-

Total Cost of Production (A+B) – 25,850/-

1. Average Output of onion per acre (in Kg's) – 4800kg's
2. Gross Returns – 76,800/- (16rs per kg)
3. Cost of production – 25,850/-

Therefore, the cost of seed is the major portion of the total cost of production of onion.

Net Return or profit = Gross Returns – Cost of production = 50,950/-

Ways to make the vegetable garden into a profitable business

Selling vegetables directly to the customer – One way to make money from growing vegetables is to sell your products directly to customers at a local farmer's market. It is a good idea to shop around to multiple different markets the year before you want to begin vegetable selling.

Connect with the customers – Be friendly to your customers and tell them stories about how your products are farmed and show pictures of the farm in order to connect them to the growing process.

Set up a roadside vegetable stand – If your farm is located on a fairly busy road in the country you could want to try selling vegetables at a roadside stand. By selling at a roadside stand you can save on transportation costs and sell vegetables after being harvested. Familiarize yourself with any government regulations surrounding the sale of vegetables at a roadside stand.

Talk to local grocery stores – Most local grocery stores, even if they are part of a big chain brand, will entertain the idea of selling locally grown vegetables produce in their store. Set up an appointment with the manager to discuss ways you can sell your product in a local grocery store.

Crops for maximizing your vegetable growing profits

The best crops for maximizing vegetable growing profits are as follows:

Growing vegetables for profit can be a very rewarding experience, particularly if you grow them organically. Choosing the vegetables to grow for-profit mainly depends on a balance between what vegetables are the most expensive to buy in the shops, yet at the same time are the vegetables most in demand.

Some of the vegetables to grow for-profit are Tomatoes, Courgettes / Zucchini, Peas, Spring Onions / Scallions, Radishes, Runner Beans, Beetroot, Cucumbers, Pumpkins, Eggplant, Cauliflower, Kale, Okra, Cabbage, Sweet peppers, Carrots, and Onions, etc.

License for selling vegetables

You should apply for the FSSAI (Food Safety and Standard Authority of India) License, whatever food and beverage product are you dealing with. FSSAI Registration is mandatory for every individual or entity that is eligible for an FSSAI food safety License.

The main step for FSSAI Food safety Registration and Licensing is identifying the nature of the business structure. When you're selling products from home, it's very important to handle food safely, even when the state leaves you unregulated. If negligence gives your customers a case of food poisoning, it can hurt business and land you in legal trouble. Your state or county agricultural department must be able to provide you with guidance on selling products from home safely. For example, suggestions for small farms;

- If you use manure as fertilizer, be careful not to contaminate your vegetable crops.
- When you irrigate crops, test to confirm the water's not contaminated.
- Train any employees you have in hygiene, food safety, and hand-washing and also follow the rules yourself.
- Make sure there's water for washing and cleaning readily obtainable when you work in the yard.
- Don't allow pets or wild animals to wander into your vegetable crops.
- Keep all the equipment and storage bins you use clean and free of contamination.

Importance of FSSAI license registration

The importance of Food Safety and Standard Authority of India is maintaining the food quality levels in order to ensure safety and providing satisfaction to every consumer is the aim of every Food Business Operator. FSSAI Registration plays the main role in formulating the controlling procedures.

The conclusion to make money from a vegetable farming business

Well, with a proper and unique vegetable gardening ideas and business plan, you can make huge money from the vegetable farming business in India.

To Start and Profit from Vegetable Farming Business

Vegetables crops and leaves are one of the major daily meals, starting from our salads, soup and other nutritious meal that are healthy.

It is highly demanded by all because of high nutritional value. It is one of the profitable farming businesses you can imagine, that grows almost all the time and grows very well as well, which that you have something to sell every time.

This vegetable farming business requires just a little capital to start and you can gain a larger profit from it, so if you are planning to start up the business, you should make sure to carry out all the market surveys and feasibility study. The tips below will help you to know how you start and make from the vegetable farming business.

1. Choose the best Species

It is advisable to always check the kind of vegetables you will like to venture into. While looking for the one you want to venture into, it is important to consider the ones that can easily be sold out around your locality. Vegetables can include artichoke, eggplant, asparagus, broccoli, cabbage, Spanish vegetable, okro known as ladies' fingers or ochro and scientifically known as Abelmoschus esculentus, green vegetable, Cauliflower, etc.

- There are various types of fruits and vegetables to select from; deciding on the right one will lead to a successful vegetable farming business.
- You should try and find out about all these around your area like; is there a high demand for this crop? How resistant is the crop to diseases and pests, the risks of planting the crop and the quantity of crop to produce first?
- Then you can decide on the type of vegetable farming you want to engage on.

2. Get a Suitable Farmland

- Finding suitable farmland is very essential when starting this kind of farming business, you should also try and know about different kinds of soil and its properties, in order to choose the best soil suitable for planting.
- It is important to consider the following while choosing land for your vegetable farming; which includes its soil type, the topography of the land, demographic, its access to sunlight and water.
- Loamy or humus soils is regarded to be the best soil type for vegetable planting, due to high nutrients present in it that helps the proper growth of vegetables.
- This type of farming can also be planted at home if you have enough space for it. But you can a small or large area of land to start up the farming business, depending on your scale of operation.

3. Get the Necessary Equipment

- Getting the necessary equipment for your vegetable farming depends on the scale of your operation.
- Your scale of operation will determine whether you need to get farming tools and machines such as planters, tractors, manure spreader, capable bakkies, etc.

- Farming equipment helps to increase the productivity of your farm. If you are planning a small scale of operation, you may consider the following farming tools such as hoe, plow, including other garden equipment.

4. Get your Start-Up Capital

- There are some ways you can raise up your start-up capital. But if you have a very small amount of funds, you can start it on a low key and expand it when you start making profits. If you are planning to start it in a bigger way, then you apply for a loan from the bank.
- Getting a suitable land for this farming business will take a lot of expenses, the seeds are also expensive and you will also need to buy some additional materials like pesticides, fertilizers, watering systems and others.

5. Pest and Diseases Control

- Vegetable crops can also be affected by pests and diseases. Therefore it is important to check on your crops for pests and diseases.
- Applying a pest and diseases control depends on the level of it has affected or the level of its damage to the crop.
- You can check on the crop to know if there is any presence of diseases and pests, by setting a trap for a particular pest and then check on it regularly.
- Some of the traps you can use are Pheromone trap (a hormone that attracts male insects) serves as a trap for pests like a diamond back moth, and Sticky yellow traps that serve as a trap for insects like aphids or thrips. Traps help the grower to know the presence and indication levels of the pest on the farm.
- Pest and disease control can be done with the use of resistant varieties, including biological controls, cultivation, trap cropping, chemical application, management and cultural methods.

6. Harvesting

Harvesting your vegetable crops should be one of the things to consider while choosing to grow vegetables. It may require hiring laborers, though depends on the type of vegetable crop you decide to plant.

Your vegetable crop can be harvested with the following means:

Harvesting by machinery: harvesting your crop with a machine does not require many laborers but the takes too much capital for the machinery, such crops include corn, carrots, processing tomatoes, and potatoes.

Harvesting by hand: this demands few laborers, though it can be managed by your family members depending on the size of your operation, and the perish ability of the crop should be considered and how fast it needs to be harvested after it matures, an example of such crops are pumpkins.

Some crops require pruning and harvesting by hand continuously and this type of crop may yield for a long time, examples of such crops are squash, snow peas, eggplant and asparagus.

Vegetable crops growing, planting and harvesting are not so difficult compared to other food crops. It is important to have your buyers ready to buy the vegetables once it has been harvested. Vegetables are perishable and storing it for a very long time can affect its quality.

7. Climate

- Climate variability is one of the things to consider in vegetable planting. It is highly beyond control, unless if you are planning to create protected cropping like shade house production or glasshouse.
- Climate will help you to know the crops that can be grown and the best time of the year to grow it. Some crops is likely to be frost sensitive while others are likely to require heat or lower soil temperature before the seeds can germinate.
- It is important to know that some specific varieties of vegetable crops like lettuce or cauliflowers are grown on specific seasonal times and conditions.
- The timing of planting some varieties of vegetable are specific, but if they are planted when it is not supposed due to bad timing or unseasonal weather it will result in not growing properly.

8. Chemical Control

The amount of chemical required for vegetable farming depends on the levels of pest present in it. Some organism is great resistance to some chemicals, therefore it is necessary to reduce the development of the resistance.

Vegetable farming has various resistance plans that have been created. Chemical labels help to provides information on how to reduce resistance. And this includes reducing the number of chemicals applied to a crop and rotating chemical groups.

There are many legal things you need to know when using chemicals, which includes sticking to the length of time after applying and when it will be harvested, having the proper chemical user certification and also the rules on record keeping.

Vegetable farming is one of the easiest farming businesses and it's very profitable. Once you have the knowledge and basic skills required, you can start the farming business even while having another job by making use of spare time to plant it and sell to retailers.

As an aspiring entrepreneur, you can decide to start your vegetable farming business on a small or large scale business, depending o your financial status.

Vegetable Weed Control Methods – A Full Guide

Weed control includes several techniques used to limit weed infestations and minimize competition. These techniques attempt to achieve a balance between the cost of control and crop yield loss, but weed control methods are used only after the problem exists; it is not prevention. Weed control techniques have been adopted widely because control is the easiest thing to do and is generally effective. What are we waiting for? let us get into the details of vegetable weed control methods.

A step by step guide to vegetable weed control methods

Weed control is very important in successful lovage production. Early weed control is particularly critical. Cultivation is an effective control technique for weeds in lovage, especially young plants. Weed control is generally performed by cultivating between the rows.

Weed control and weed management are the two terms that are used in weed science. Weed control is the method of limiting the infestation of the weed plant so that crops can be grown profitably. While, weed management includes prevention, restricting invasion, suppression of growth, prevention of seed production and complete destruction. Therefore, weed control is one of the aspects of weed management.

The critical success factors for weed control include;

- Choose the suitable methods by considering the available herbicide options, preparing beds to reduce the weed seed bank, as well as the role of tillage, cover crops and crop rotation within your system.
- Consider timing for example weather and access to paddocks, weed life stage, crop life cycle, and ground conditions, particularly moisture.
- Be diligent about applying suitable weed control methods, focusing on removing weeds before they set seed or spread further.
- Have good knowledge of main weeds and their characteristics that are trial herbicides on-farm.
- Develop a plan, and implement an Integrated Weed Management (IWM) that combines a number of control methods and looks after resistance management (covered in the next section).

Cultural methods of vegetable weed control

Several cultural practices like tillage, planting, fertilizer application, and irrigation, etc., are employed for creating favorable conditions for the crop. These practices if used correctly help in controlling weeds. Cultural methods, alone cannot control weeds, but mainly help in reducing the weed population.

Cultural weed control refers to any method that involves maintaining field conditions such that weeds are less likely to become established and increase in number. Examples of cultural weed control in vegetable crops would be crop rotation, avoiding overgrazing of pastures or rangeland, using well-adapted competitive forage species, and maintaining good soil fertility.

Preventative vegetable weed control methods

Preventative weed control refers to any control process that aims to prevent weeds from being established in a cultivated crop, a pasture, or a greenhouse. Some examples of preventative weed control would be using certified weed-free seed, only transporting hay that is weed-free, making sure farm equipment is cleaned before moving from one location to another location, and screening irrigation water to prevent weed seeds from traveling along irrigation ditches.

Chemical weed control in vegetable production

Herbicide application can give the most effective and time-efficient method of managing weeds. Numerous herbicides are obtainable that provide effective weed control and are selective so that grasses are not injured. Along with herbicide use, user responsibility and compliance with all product label conditions for herbicide handling, use and cleanup are required.

Chemical weed control refers to any method that involves the application of a chemical (herbicide) to weeds or soil to control the germination or growth of the weed species. In economic terms, chemical control of weeds is a large industry and there are scores of examples of chemical weed control products.

Chemical control in vegetable seedbeds

There are fewer registered herbicides for seedbeds than for planting crops. There are some post-emergence grass-killers (usually known as 'fop' and 'dim' families) that could be used well in vegetable seedbeds, as for example, cycloxydim (for onion, cruciferous crops), clethodim (onion, tomatoes), fluazifop-butyl (tomato, pepper, lettuce, leek, onion). Rates should be low to avoid any problem of phytotoxicity.

Herbicide treatments under plastic cover are always hazardous and careful application must be carried out. Under plastic, high levels of moisture and elevated temperature are common and plants grow gently. Selectivity could be simply lost and phytotoxicity symptoms may occur, while sometimes they are just temporary. The effects are often erratic and the best way to deal with it is to be prudent and make some trials before a general treatment.

Biological weed control of vegetables

The goal of biological control in vegetable crops is not eradication but the use of living agents to suppress vigor and spread of weeds. Such agents can be insects, bacteria, fungi or grazing animals, for example, sheep, goats, cattle or horses. Grazing produces results similar to mowing and biological control is most commonly thought of as "insect biocontrol".

Biological weed control mainly refers to any technique that involves the use of natural enemies of weed plants to control the germination of weed seeds or to control the spread of established plants. This is a rapidly expanding area of weed control with several examples. Examples of biological weed control contain sheep to control tansy ragwort or leafy spurge, cinnabar moth and the tansy flea beetle to control tansy ragwort.

Mechanical weed control of vegetable farming

Mechanical weed control consists of methods that kill or suppress weeds through physical disruption. Such methods contain pulling, digging, disking, plowing and mowing. The success of several mechanical control methods is dependent on the life cycle of the target weed species.

Mechanical weed control mainly refers to any technique that involves the use of farm equipment to control weeds. The two mechanical weed control techniques most often used are tillage and mowing.

The critical period of weed control in vegetable field

This period has been defined as an interval in the life-cycle of the crop when it should be kept weed-free to prevent yield loss. Horticultural crops are sensitive to weed competition and need to keep them weed-free, from planting, emergence or until the end of their critical weed-free period. If the crop is kept weed-free for the critical period, normally no yield reduction would be there. Again, weeds emerging after the critical weed-free period will not affect crop yield. But control efforts after this time can make harvest more efficient, or reduce weed problems in subsequent years.

The critical period of weed competition is generally longer in direct-seeded than in transplanted crops. Some traditional techniques that are transplanted, earthing-up as done in potato are thought to increase crop competitiveness. Weather conditions and weed density have an influence on the length of critical periods.

Weed control methods in vegetables

Weed control in vegetables particularly important early in the season when weed competition can substantially reduce vigor, uniformity and overall yield. Vegetable crops are very sensitive to weed competition and need to keep them weed-free, from planting, emergence or until the end of their critical weed-free period. The period from emergence to four weeks has been found to be critical in the competition of weeds in several row crops including vegetables.

Only a few vegetables are excellent competitors with weed flora because they quickly cover the soil, topping the weed growth like potato, transplanted brinjal, and cabbage. Some vegetables, such as carrots, turmeric or direct-seeded vegetable crops like cabbage produce slowly and they cover the soil very sparsely, suffering strong weed competition not only for water, nutrients, and light but even for space. Therefore, if weed control is not carried out timely, there will be no production at all. There are several examples of problems in crop-yield reduction that indicate the great sensibility of vegetables to early weed competition and the need to control weeds at early crop stages. Weed competition is severe when a direct-seeded vegetable is grown.

Some of the weed control methods in vegetables are given below:

Crop competition

Crop competition is one of the cheapest and useful methods a vegetable producer can use to control weeds. Competition makes full use of one of the oldest laws of nature and survival of the fittest. Some of the competitive crops are closely planted bush green bean, cucumber, winter squash, and sweet potato. Some vegetables sweet corn, pole bean, watermelon and transplanted tomato, pepper, and cabbage are intermediate. Slow or short growing crops for example as carrot, onion, okra, and direct-seeded tomato, pepper, and cabbage are considered non-competitive.

Weeds are naturally very strong competitors. If not, they would fail nature's test of "survival of the fittest" and those weeds that can best compete always tend to dominate. For example, some weeds germinate and grow quickly and may dominate a field seeded with a slower-growing vegetable.

Weeds compete with crops for light, soil moisture and nutrients, carbon dioxide, and physical space. One wild mustard plant could consume twice as much nitrogen and phosphorus, four times as much potassium and four times as much water as a well-developed onion plant. The average ragweed consumes almost three times as much water as does a sweet corn plant. Weeds, as a group, have much the same requirements for vegetable growth. For every pound of weed development, the soil produces about one pound less of crops. The competition for nutrients can be simply seen from chemical analysis of weeds growing and competing with corn.

Crop rotation method for weed control in vegetables

Certain weeds are more common in some vegetable crops than in others. This is due to the time of year that the crop is grown. For example, pigweed, lambs quarters, common ragweed and crabgrass often are mainly found in tomatoes, peppers, vine crops, sweet corn and in other summer cultivated crops. Wild mustard, volunteer small grains, wild garlic, cornflower, and thistles are weeds that happen frequently in the fall and winter vegetables.

Rotation of vegetable crops can be an efficient method to reduce weed populations. A good rotation for weed control generally includes strong competitive crops grown in each part of the rotation.

Biological predators and diseases

In biological weed control, a "natural enemy" of the plant is used which is harmless to preferred plants. Insects or disease organisms generally are natural enemies. Parasitic plants, selective grazing by livestock and rodents, highly competitive replacement plants are other forms of biological control.

Seedbeds

Many vegetables are grown in seedbeds to develop appropriate seedlings for transplanting in the field. Soils dedicated to seedbeds are generally light, with good tilth, and fertilized to obtain a good plant emergence. Seedbeds are generally flood-irrigated and plastic-protected. Here we add some possibilities for weed management.

Stale seedbeds

Stale seedbeds are sometimes used for vegetables when selective weed-control practices are limited or unavailable. Mainly, this technique consists of the following points;

- Preparation of a seedbed 2 to 3 weeks before planting to achieve maximum weed-seed germination near the soil surface.
- Planting the vegetable crop with minimum soil disturbance to avoid exposing new weed seed to favorable germination conditions.
- Treating the field with a non-residual herbicide to kill germinated weeds just before or after planting, but before crop emergence.

Soil solarisation of weed control method

Soil solarisation method is a broad-spectrum control method, which is simple, economically feasible and environmentally friendly. It is an effective way for the control of many weeds. It does not affect soil properties and generally produces higher yields. There are some disadvantages to its implementation. For example, previous irrigation is a requirement, (or frequent and abundant rain) and the soil should be kept solarized (non-producing) for a period of at least one month. Results are often variable, they depend on weather conditions. Cold (high latitude) or cloudy places are usually not appropriate for implementing solarization. Some species can tolerate solarization (e.g. deep-rooted perennials are Sorghum halepense, Cyperus rotundus, Equisetum spp. and also some big weed seeds for example legumes).

The soil should be clean, surface-leveled and wet, previously to being covered with a thin (0.1-0.2 mm) transparent plastic sheet and very well sealed. The soil should be kept covered during the warmer and sunnier months (30-45 days). Soil temperatures should reach above 40° C to exert a good effect on weed seeds. After solarization, the plastic should be recovered, and the use of deep or moldboard tillage must be avoided. This system is suitable for small areas of vegetables, but it has been mechanized for extensive areas of tomatoes. Soil solarization process is widely used under plastic greenhouse conditions. Hope you enjoyed the content of vegetable weed control methods.

3

GINGER FARMING AND PRODUCTION MANAGEMENT

Vintage people of India 'makaolacata" articulated by the world's leading medical substances Ginger is used in Chinese and Japanese people for a long history of medical substances containing aromatic substance. To this day, the traditional medical systems in the world occur during pregnancy, vomiting, dizziness and cold is used as the primary clinical. Medicinal ginger rich, the people of India, China and Japan in the cooking methods ginger is essential. Thus, all hot and temperate regions of the world in medical popular ginger have been widely cultivated. South East Asian Countries are the origin of Ginger.

From India and China, the crop spread to other countries. Currently, Ginger cultivated in countries like India, China, Jamaica, Taiwan, Fiji, Mauritius, Brazil, Indonesia, Japan, Malaysia, Philippines, Thailand and Bangladesh. It is in the Gingiferae plant family and 'Gingifer officinali is the botanical name. The Latin name of the genus cinjiper was derived from the Tamil word injiver. The name Ginger spread by Arab traders to the Greek and Roman developed countries.

Ginger scientific name is *Zingiber officinale* and belongs to the family Zingiberaceae. It is an herbaceous perennial plant, where its rhizomes are used as a spice. India one of the leading producer of Ginger in the world. Apart from spice it also has many medicinal values, it is considered as the healthiest and delicious spice. It contains many nutrients and bioactive compounds that have powerful benefits for your body and brain.

The following are the proven health benefits supported by scientific research-

1. Gingerol a substance in ginger has powerful medicinal properties.
2. Ginger is the best medicine to treat all forms of nausea mainly Morning sickness.
3. Ginger helps in reducing muscle pains and soreness.

4. It has an excellent anti-inflammatory effects hence can be used in treating Osteoarthritis.
5. Recent studies proves that ginger is effective in lowering the blood sugar levels.
6. Ginger helps well in treating the chronic indigestion.
7. Ginger powder helps in reducing the menstrual pains.
8. Ginger lowers the cholesterol levels.
9. Ginger also helps in preventing cancer.
10. Ginger is effective in brain function and hence helps in protecting against Alzheimer's.
11. It also effective in fighting against the infectious diseases.

What are we waiting for? let us see the cost and ginger farming income per acre.

Production Status

It cultivated by all tropical and subtropical countries in the world, much of it is produced in India. By almost 30 percent to 40 percent of world production is produced in India. In India, states like Kerala, Arunachal Pradesh, Orissa, Meghalaya, West Bengal are widely cultivated and, Tamil Nadu, Andhra Pradesh and Karnataka are cultivated in the lower area. In Tamil Nadu highly rain fall districts like Kanyakumari and lower palani hills where the ginger cultivated.

Benefits

India, China and the Japanese people are used ginger as food and all the people used ginger as medical substances. Colds, cough, vomiting, dizziness, hypertension, impaired vision problems, such as the ginger is used as drug. Indian Ayurvedic treatment for digestion, fever and stomach diseases, ginger is used heavilyin the production of medicines.

Chin called the dry ginger and pepper, along tippili tea helps to removing mucus in the body for a long time and helps to protect health. Aromatic and medicinal oil extracted from the ginger, helps in preparation of ginger beer, ginger wine. Ginger powder used in the spices products. Foreign countries, ginger used for making Biscuits, Cakes, soup and pickle.

Cultivation practices

Climate

It grows well in medium temperature regions and moisture in the air. Ginger cultivated 1500 meters above sea level. But it grows well in 300 meters to 900 meters above the sea level. At regular intervals throughout the year from 1500 to 3000 mm of rain per year available rainfall areas should be chosen. Land preparation and harvesting periods of dry weather are needed. If cultivated in areas where rainfall is less watering should be done at regular intervals.

Soil

Clay sediment organic content of the soil is packed with all kinds of land area, ginger can able to grow. But choose the land by at least 30 cm deep in the soil is necessary. Alluvial gravel mixed red soil is also suitable for ginger cultivation. In 1 meter deep, well-drained soil the ginger yield is high.

Soil pH should be 6 to 6.5. Drained lands should be avoided. Ginger well absorb the nutrients in the soil. By their nature, avoid ginger cultivation again and again in the same field.

Propagation and seed rate

Ginger is propagated by seed tubers. Seed tubers that has 5.0 to 2.5 cm in length and weigh from 20 to 25 grams are used to make one or two viable nodes are used in propagation.

1500 to 1800 kg/ha seed tubers need for plains and hilly areas 2000 to 2500 kg/ha are recommended.

Those that used as seed tuber, after the harvest, it should be separated according to the quality. Select the seed tuber without disease infection and dip it into the 0.3% Mancoseb and 0.1% Malathion Solution and dry it in shade, should be stored using a sand pit. Rot checked once a month, if any, should be cleared.

Planting season

Ginger is a rainfed crop. It cultivated dependent on the rainfall. Thus, the crop planting should be done in the first week of April. In west coast, ginger planted month of May after a summer rain. But in places with good irrigation facilities and can be planted until mid to late February and March.

Land preparation

After the summer rains, plough the land as fine tilth. In deep well ploughed field raise 1 meter width, 15cm height and enough length beds at 50 cm intervals. If irrigation facilities available raised a bar with 40 cm intervals.

In Nematode and root rot disease affected areas, 40 days before planting, should steamed soil cover beds with polythene. By doing this, the cultivation can be done without damage.

Seed treatment

Seed tubers should be soaking by 5g Pseudomonas 10 days before sowing and dried in the shade. By doing this, not only to increase visibility of dormancy and germination of the nodes can protect from disease.

Sowing

Before sowing the seeds, 25 -30/ha FYM and 4 kg/ha Pseudomonas spread mixture on top of the beds. If spread 2t/ha neem cake powder, it protect from root rot disease. Then seed can be sown at a spacing of 20-25 cm. When planting a maximum depth of 5 cm should be transplanted. After planting, sand mulch should be done.

From the date of sowing, depending on soil moisture will begin rising from 25 to 35 day.

Application

When planting to harvest good yields by the above basal levels of manure and neem cake powder to be applied. Recommended dose of NPK is 75:50:50 kg/ha. On this 50% of the phosphorous should be applied as basal. Nitrogen and potassium should be applied 50% (37.5 and 25kg) after 40 days of planting and remaining dose should be applied after 90 days of planting. Boron and 3 kg / ha and zinc 5 kg / ha of mineral fertilizers in crop yield by providing additional research results suggest.

Mulching

Laying mulch increases seed germination. By increasing soil moisture and protects the organic nature soil mulch setting considered necessary. Mulching will be set during the first planting. 10-12 tons/ha green leaf or 5-6 ton/ha dried leafs are used for mulching. Similarly, 45 and 90 days after planting, mulching is to be set. Easily available with coconut leaves, leafy branches, banana leaves, rice paddies, coconut pith, sugarcane trash of the inputs can be used as mulch.

Water Management

To obtain good yields in ginger 1320 to 1520mm rainfall is required. Planted in April-May, depending on the moisture of the soil by watering 2-4 times a week for the time required. In the absence of rain, watering should be done at intervals of 15 days. Seed tuber germination time in ginger and tubers produced time moment essentially require watering times.

Weed management

Weed management is carried out twice during the cultivation of ginger. Weed removal of the first to set up Mulching, the second for the second time before the weeds must be removed at intervals of 45-60 days, depending on the amount of weed. When we remove the weed, the plant needs to be taken care of without affecting stem and root. After completing the weeding the earthih up should be done.

Crop rotation

Ginger absorbs the large amount of nutrients in the soil. So ginger should not be cultivated the same land continuously. So it should be cultivated with rotation of tapioca, beans, chow-chow and other vegetables. Ginger can be cultivated as a single crop or intercrop with coffee, orange, banana etc.

Plant Protection

Insects

Stem borer

Stem borer causes the most damage in ginger. If the insect infested plants become yellow color of the leaves, stems drying. To control pest, a low-level attack subjecting Peo - de – Janeiro may be used for cultivation. It should be control using prakkonite family of natural enemies or spray the pesticide monocrotophos 0.1% (1 liter of water 1 ml) and infested plants should be removed.

Leaf roller

Leaf roller attacks the leaves and leave are rolled, which are found in large numbers in the months of August and September. To control it, to spray carbaryl 0.1% (1g per liter of water) or dimethoate 0.05% or phosphamidon 0.05%.

Tuber scales

Tuber scales attacks the tuber in the field and also in godowns. It attack the plant severely, so the plant dried. In godown tuber scale buds dried and it reduces the germination capacity. To control the tuber scales, thrips and other sucking insects, spray phosphamidon 0.05 per cent (5 ml with 10 liters of water). Spray malathion in the bag which is filled with tuber.

Diseases

Tuber rot disease

It is the most important disease which attacks the ginger. Where there is not even adequate drainage areas of stagnant water found in this disease. The disease is transmitted through the soil and seed tubers To control the disease, improve the drainage facilities. Use disease free tuber for germination and the tubers are treated with Mancozeb or copper oxychloride, a fungicide, a liter of water, 3 g of the quantities produced in the solution for 30 seconds or before saving the tuber soak with streptocakki 200 ppm (ie, a liter of water, 200 ml). By placing large quantities of farmyard manure along with a decrease of the disease seriously.

Leaf spot disease

To control the disease, Bordeaux mixture 1% or copper oxychloride 0.25% (2.5 g per liter) apply as a foliar spray.

Microbial wilt

To control the disease, the tuber can be treated in steptocycilin 200 ppm before sowing.

Ginger Farming Income; Cost; Profit; Project Report

Ginger is cultivated in many of the states in India however states namely Karnataka, Orissa, Assam, Arunachal Pradesh, Gujarat, and Meghalaya contributes about 65% in India's total production.

Climate and Soil conditions required for production of Ginger:

Ginger grows well in warm and humid climates. It can be cultivated in rain fed and irrigated conditions. But it requires a moderate rainfall at sowing time until the rhizome sprouts, during dry weather and growing seasons well distributed showers are required, and also a month before harvesting. Ginger grows well in drained soils like sandy loams, clay loams, and red loams. It grows best in soil rich in humus.

Ginger varieties:

These are the prominent high yielding varieties of Ginger- Maran, Kuruppampadi, Erad, Wayanand, Himchal and Nadia, exotic variety Rio-de-Janeiro is the most popular variety among cultivators. The IISR Varada variety is used for fresh ginger, dry ginger and also in making candy, whereas IISR Rejatha is rich in essential oil.

Planting season:

It can be planted under well irrigated conditions; hence it can be planted during the months of February or early March. This planting has given more yield and also showed resistance towards diseases. The other type of planting used in West coast India is during the first fortnight of May with the receipt of pre-monsoon showers.

Land preparation: The land has to be ploughed thoroughly and deeply nearly 4 to 5 times with receipt of showers to bring the fine tilth to soil. Beds of 1 m width, 30 cm height with minimum length are prepared and an interspacing of 60 cm in between the beds; this has to be maintained for plantation during May, whereas for the irrigated land 50cm ridges are formed.

Note: If the land is prone to rhizome rot disease or nematode infestation solarization of beds for about 40-45 days using transparent polythene sheets is the best method.

Planting of Ginger

Ginger is propagated through seed rhizomes, which are cut into small pieces 2.5-5.0 cm length and weight of 22-25g each, whereas every piece should have two or more good buds. The cost of rhizome seed varies from place to place along with the method of cultivation adopted. The seed rate varies from 600-750/Kg/acre. Ginger yields better if the seed rhizomes are treated with 0.3% mancozeb for 30 min and dried in shade for 4hrs. These treated seedlings have to be planted with a spacing of 20-25cms along the roads and the same in b between the rows; and has to be covered with well decomposed farmyard manure along with a thin layer of soil.

Manures and fertilizers

Well decomposed cow dung or cattle manure or compost has to be applied at the time of planting at a rate of 10-12tonnes/acre. Application of neem cake helps to reduce Rhizome rot disease or nematode and also increases the yield. Recommended dosage of N:P:K is 200:75:100. Zinc Fertilizer of 2 kg per acre gives good yield if the soil is deficient in Zinc Micronutrient.

Irrigation

Ginger can be cultivated as rain fed crop in high rainfall areas where as irrigated crop in low rainfall areas. The first irrigation has to be done immediately after planting. From then irrigation is given once in 10 days based ion the weather and soil type. Drip system and Sprinklers can also be used for better efficiency and to increase the yield.

Ginger plant protection

Diseases such as soft rot, bacterial wilt, leave spot and nematode pest can be controlled by using neem cake, Bordeaux mixture, Mancozeb, Copper oxychloride.

Insect pests such as Shoot borer, rhizome scale, minor pest can be controlled by spraying Mallathion, Quinalphos, and Chloropyriphos.

How and when to harvest Ginger

Ginger takes 210-240 days to mature completely (dry ginger) after planting. But based on the demand it can be harvested after 180 days. For making dry ginger the rhizomes are harvested after full maturity that is when the leaves turn yellow and start drying. Irrigation has to be stopped before 1 month of harvesting. In wide space cultivation, tractor or power tilter harvesters are used. Soil adhering the manure has to be separated manually. Based on the demand, late harvesting is also under practice but in India fresh green ginger is mostly preferred and the bleached and unbleached dried ginger are in demand to export to other countries.

The yield of ginger per acre:

The average yield of green ginger is 6-10 tonnes/acre whereas the dry ginger varies from 16-25%.

Cost for Cultivation of Ginger herb:

- Cost of Seedlings: Rs.35,000 per acre
- Cost of FYM: Rs. 6,000 per acres
- Cost of Fertilizers: Rs. 5,000 per acre
- Irrigation charges: Rs. 8,000 per acre
- Plant Protection charges: Rs. 3,000 per acre
- Labour Charges: Rs. 3,500 per acre
- Transporting charges: Rs. 1,500
- Miscellaneous charges: Rs. 1,000
- 10% of total cost :Rs. 12,150
- Total cost incurred : Rs.1,33,650

Income returns:

- Cost per Kg =Rs. 35 as on September 5th 2019.
- Note: This rate differs region to region and place to place
- So, income for 8 tonnes = 8,000 x 32= Rs. 2,56,000
- Profit = Income – Total cost= Rs. 2,56,000- Rs. 1,33,650= Rs. 1,22,350.

Conclusion of Ginger farming income:

The net income for cultivation of ginger in 1 acre is Rs. 1,22,350.

Harvest

Ginger plant preparing for harvesting within eight months from the date of planting. Brown leaves are ready to be harvested at arrival points to the bottom-up dries. This condition should be harvested in order to extract the ginger oil. The leaves are harvested in the dry state by seed tuber used. Harvested ginger, removed from the dried leaves, roots, and tubers, unearthed by the adhering soil is removed, washed with water and dried in shade. The ginger used as vegetable and cooking, it should be harvested from the fifth month after sowing. This immature harvested ginger has less alkalinity and fiber.

Yield

Well maintained crop average yield from 15 to 20 tons/ha of ginger rhizomes. Post harvest

Ginger - and sell it off as soon as ginger, washed, dried in shade and then stored for 2 days. If we keep the ginger in 55 cm heat and 65% of humidity, stored upto 6 months.

Dry ginger - used after drying.

Black Sin - the rhizomes are put in boiling water for 10-15 minutes, remove the skin and dried. It is available in black color is killing tubers.

Processed ginger

Processing in sugar solution: Ginger rhizomes should be well washed, cut the roots and then cooked in sugar solution and then dried.

Processing in salt water: wash the rhizome, peel skin, soak it in salt water. Then shall pour the water mixed with a little salt and vinegar and soak in 7 days. Then rinse well, and sterilize with 10 minutes. Drilled the tuber and soak the sugar solution (48 kg sugar and 60 kg of ginger) in the quantities to be sterilized. Boiled it three times and dried it.

Ginger Candy

Cook the tuber well and drained it and then kept in the syrup (for syrup 1 kg to 1.5 kg sugar, add ginger) and boil it for two hours and to keep cool. Pour the syrup over the top of the plates and cover it. It could taste the longer lights up. The material is made from ginger is ginger oil and oliyoresin.

Seed tuber storage

Increase the seed tuber germination capacity must be stored in pits on the spot in the shade. Choose the tubers for seed without infection. For this choose the plant of six to eight months of age without disease infection. Seed tubers should be treated with Quinalphos 0.075% and Mancozeb 0.3% mixture, soak 30 minutes in the solution. Apply cow dung on the lateral wall of the storage pit. Arrange the seed tuber in several layers in the pit. After every layer broadcast the sawdust for 2cm. Top of the pit with the flow of air give a little gap and close with wood. We should check the seed tubers for three weeks interval and eliminate the shrunken and disease infected tubers.

Seed tuber storage

December – January month harvested tubers are sowing in the month of May – June. So we have to store the tubers for 3 to 4 months. Seed tubers traditionally stored in sand or pits in shade. Because of this the tuber dry and rot disease affected before sowing and germination, post harvest loss is very high. Seed storage as a result of research in technology-ventilated seed rhizome 3 percent to 100 gauge polythene bags, tying it down on dry sand saving, post harvest losses are decreased viability in seed tubers found. Soak the seed tubers in Quinalphos 0.1% or Dythene M 45 and dry it and store the tubers.

Intercultural operations

Mulching

Mulching the ginger beds with green leaves is essential to enhance germination and to prevent washing off soil due to heavy rain.

Weeding

Weeding is done just before fertilizer application and mulching. Two or three weedings are required depending on the intensity of weed growth.

Crop rotation and mixed cropping

Ginger is commonly rotated with other crops. The crops most commonly rotated with ginger are tapioca, chillies, dry paddy in rainfed areas and ragi, groundnut, maize and vegetables, red gram and castor.

Plant Protection

Insect pest

Shoot borer

The presence or bore hole on the pseudo stems through which the frass is extruded and the withered central shoot are the symptoms of pest infestation.

Control

Spraying of Malathion 0.1% during July-October at monthly intervals is effective against the pest.

Leaf roller

The larvae of the leaf roller cut and fold the leaves and remain within and feed on them.

Control

In severe infestation, spraying with carbaryl 10-1% or dimethoate 0.05% may be undertaken.

Diseases

Soft rot or rhizome rot

The disease is soil borne and the fungus multiplies with build up of soil moisture with the onset of south-west monsoon. The collar region of affected pseudostem becomes water soaked and rotting spreads to the rhizome resulting in typical soft rot.

Control

Treat the seed rhizomes with 0.3% Dithane M-45 for 30 min. Drench the beds with Dithane M-45 (0.3%) or Captofol (0.3%).

Bacterial wilt

This is also a soil and seed borne disease. Water soaked spots appear at the collar region of the pseudostem and progress upwards and downwards. The affected pseudostem or the rhizome when pressed gently exudes milky ooze from the vascular strands.

Control

Seed rhizomes may be treated with 200 ppm Streptocycline for 30 minuted and shade dried before planting. A uniform drenching be given to all beds with 1% bordeaux mixture or 0.2% copper oxycchloride.

Preservation of seed rhizomes

For seed materials, big and healthy rhizomes from disease free plants are selected immediately after the harvest. For this purpose, healthy and disease-free clumps are marked in the field when the crop is 6-8months old and still green. The seed rhizomes are treated with a solution containing 0.1% quinalphos and 0.3% Dithane-M-45 for 30 min. Drain the solution and dry the rhizomes under shade. The seed rhizomes are stored in pits of convenient size in sheds.

Dry ginger

For dry ginger the produce is kept soaked in water overnight. The rhizomes are then rubbed well to clean them. The rhizomes are removed from the water and the outer skin is removed with bamboo splinters. The peeled rhizomes are washed and dried in sun uniformly for one week. To get good appearance, peeled rhizomes are soaked in 2% lime water for 6 hours and dried thereafter. The yield of dry ginger is 16-25% of the fresh ginger depending on the variety.

4

FRUIT FARMING BUSINESS

Fruits are among the constant commodities in the food market. Almost everybody eats fruits due to their nutritional value, or simply because they are delicious. Accordingly, growing fruit trees can be a good source of income for every farmer and new entrepreneur desiring to tap the potential of the fruit-growing industry. With these, the following are the basic things one needs to know when starting a fruit farming business.

- Fruit farming is the growing of fruit crops which comprises of all seed-bearing structure in flowering plants as well as nuts for commercial and consumption purpose. Here, this article is completely discussing fruit farming for commercial purposes.
- Having known your sole aim of venturing into fruit farming, the next step to take to make it a success is to construct a comprehensive business plan just as provided below.
- It is of great advantage that most parts of the world are so much blessed with fertile soil which makes it good for growing diverse ranges of food crops including fruits and vegetables.
- Starting a fruit farming business is growing day by day in the world as farmers and other entrepreneurs have profited in this business venture.
- Fruits are a staple commodity that can be found in the markets just like other food crops and in numerous quality and types which generally has great health and benefits.
- Fruits and as well highly sought after and this has made it be one of the lucrative businesses anyone can engage in or use to complement their earnings.
- In fruit business, it doesn't matter if you are a professional farmer or still a novice, once you follow the guidelines below you will be able to succeed in any fruit farming business you chose.

1. Select A Particular Type of Fruit

There are up to 2000 types of fruits around the whole world and thousands of varieties as well. You cannot farm all the fruits at ones, so what you have to do at the initial stage is to select one particular type that is mostly consumed in your area.

- It is better you know what kind of fruit that you want to take as your business venture. Some of the fruits are strawberry, mangoes, oranges, and cherry to name but just a few.

The passion fruit you have been dreaming of farming may not be suitable in your area. Some important things you need to consider before choosing the fruit type no matter where you are in the world include;

- What kind of fruits the market in your area need
- What is the demand for this fruit like?
- What volume of this fruit should I start with?
- What kind of fruit can effectively grow in your region considering weather and climate conditions?
- Which one will yield more profit?
- How resistant is this fruit to pests and diseases?
- Is there any risk associated with growing this fruit?

After you must have gotten answers to your questions during the course of your market research, then you can take the next step to forge ahead with what you are out for.

2. Study The Fruit You Have Chosen

For any fruit business that you want to start, make sure that you have the basic knowledge about that fruit and how to plant it.

Also, its best planting season, how to get the seedlings, care and maintenance processes required, when to add fertilizer, how to know it's time for harvest and lots more.

You can get the proper training by working with a farmer that is basically specialized in the cultivation of such fruit in order to get the basic knowledge that will help you succeed.

Additionally, try to find out the varieties available for that particular type of fruits and which one is most profitable for commercial fruit farming.

3. Where To Start

You need farmland to plant your fruit trees, just as every farming business needs a land. The land is considered the most important element when planning to establish a farming business.

As a beginner, it is advisable to get a small farming portion instead of acquiring a large piece of land until you have successfully passed your experimental stage.

When choosing land for fruit farming business, don't just choose for farming's sake but consider soil type and composition which is one of the main determinants for effective yield and growth of plants.

The soil should be composed of humus and other essential minerals which can be determined by conducting or carrying out a soil test. However, you can also add more vital nutrients to the soil when preparing the land for cultivation by adding organic manure or fertilizer.

Check the pH value as well to make sure that it is normal, neither too acidic nor alkaline.

Suitable farmland for fruit farming should have access to sunlight exposure; topography and demography; good soil type/soil PH and availability of water sources which are all beneficial to the fruit plant. Do not neglect any of them.

4. Start With Less Quantity

The first trial of every business is always considered as the experimental farming stage.

That is; a stage of make the first demonstration fruit farming; test possible hypothesis and make a discovery that will help you to move on while avoiding possible mistakes.

Starting small will help in the effective management of the fruit crops, and will make you to easily face some challenges which might occur at the course of the farming process.

From this little starting step, you will learn from your mistakes and gain better experience that will help you handle large fruit farming operations in subsequent trials.

5. Good Starter Trees And Best Seedlings Matter

Get the best seedlings from other local farmers or from any ministry that oversees the affairs of farming and agricultural practice in the country.

Getting the right and the healthy seed will determine what you will gain in return when they fully develop and mature into fruits.

6. Plant The Fruit Trees

Fruit trees are planted at any season of the year. In a case whereby your location has hot summer or cold winter, you have to wait till spring to enable the plant to adapt to the soil.

You can buy a grafted fruit tree either in a bare-rooted form of potted tree form. The trees should be planted immediately they are purchased.

It is advisable to buy from a local nursery around the vicinity to ensure that you have got the trees that do well in that particular region.

To plant the tree in the soil, dig a hole that will be a bit wider than the plant then put directly at the center of the pit and cover the roots very well with the soil.

7. Disease And Pest Control

Fruit trees are attacked by certain diseases and pests. Therefore you should not wait till it starts before sourcing for ways to control them.

Get the right pest control such as fungicides and disease control chemicals on time and ensure to spread carefully to avoid damaging your plant or the fruits.

For effective management of your plant, you have to keep an eye always on the plant to ensure a good growing state.

8. Source For Target Market On Time

Fruits are highly perishable, so while waiting for it to grow, use that opportunity to start sourcing for new and additional customers.

Once it is time for harvest you can confidently sell to already available potential customers to avoid perishing and discouraging the effort you have made so far.

9. Harvesting & Storage

Fruits are harvested a couple of months after planting depending on the type of fruit you are farming.

Generally, they are harvested when they are full size and the normal fruit color when ripening; the rinds should be glossy and firm at this stage.

Fruits should be kept in optimal storage locations. However, fruits are not to be stored for a long period of time because they can perish easily.

What is fruit farming called?

Fruit farming in agriculture is a branch of botany called pomology which is the study of fruit cultivation; while the person that studies pomology is known as a pomologist.

How do fruits grow?

Just as fruits differ in types so the plants as well. Each type of fruit emerges from different types of plants.

Some fruits grow on trees. Examples: cherries, bananas, apples, etc. Some others grow on vines such as strawberries, grapes, and watermelons while the rest like raspberries and it's related grow on bushes.

Fruit Farming Business Plan

Every business venture needs careful planning and strategizing at the start. This plan necessarily includes all important and necessary matters to the farm. Having a definite plan will give the entrepreneur an overall outlook regarding the business, together with possible problems, which may be encountered. It involves the startup cost for the fruit farming biz, the place to start planting the orchard, the equipment necessary for farming and transporting the fruits, the market, the needs of buyers, and the existing fruit farming biz, competitors.

Where to Start a Fruit Farming Biz

One of the most important elements of every farming business is the land. Choosing the right location for your orchard greatly affects the growth and production of your business. Several factors must be considered in choosing the land to plant your orchard. This includes the type of soil, the soil's PH, exposure to sunlight, and source of water. In determining these factors, an orchard owner must first determine what type of fruits to grow on the farm. It also involves choosing the size of the trees to be planted. As a general rule, the dwarf the trees, the sooner it will produce fruits.

Equipment to Use

A fruit farmer does not need to invest much in equipment when the farm is not that big. The most common and basic tools needed are an air-blast sprayer, small tractor, mower, pull behinds, and carts.

The air-blast sprayer is for fungicide, insecticide, and application of plant growth regulator. The tractor on the other hand should be small enough to run around the trees in the orchard. A 30-60 hp tractor is sufficient for small farms. A cart or small truck is also necessary to transport the harvest from the farm to the storage area, and also to the market and other fruit farming biz clients.

Settle All Legalities

Even if your business grows over time, the need to settle all legal matters must not be taken for granted. Legal issues may arise in the future, which will hamper the growth of your business or even lead it to eventual loss. It is best to deal will legal matters at the start of the business by going directly to the local government office. Secure a fruit farming biz license. Ask for the requirements and comply with them accordingly.

In an agricultural economy like India, fruit farming proves to be a profitable business in many ways. Even though setting up your farm may cost you big bucks but the high yields of these crops, once the plants mature, in your farm will not only reimburse the cost, but you'll also end up making some good profits out of it, as fruits sell in rates higher than vegetables! Apart from that, you can grow plants in lesser spaces than other crops.

Most Profitable Fruit Farming In India

Fruit trees are typically multipurpose in that many provide not only fruit but also medicinal products and livestock fodder, with fuelwood and timber at the end of their productive lives. In addition to economic benefits, fruit trees give several environmental services.

In this article we cover all the following topics;

- Tips for growing fruits for profits
- Banana production per acre
- Mango growing season in India
- Profitable fruits in India
- Mango profit per acre
- Importance of fruit farming in India
- Factors affected by profitable fruit farming

Profitable fruit farming In India

Fruit farming is an important sector of horticulture and it has immense economic importance. This is a high paying option for land utilization. Nowadays, you will find improved fruit seeds and developed technology for fruit farming. The profitability depends mostly on demand. Your fruit farming business will be profitable if there's a demand for your product. The most profitable fruit crops have the most demand. The profitability of a tree mainly depends on location, demand, environment, and ease of growing the tree.

Tropical and sub-tropical fruit trees are generally evergreen and perennial and are frost sensitive with little growth below 10°C. Tropical species are distinct from subtropical species in that they need humid conditions and are sensitive to temperatures below 20°C. They thrive in climates where average mean temperature ranges are higher than 10°C for the coldest month. The widely cultivated and globally traded tropical and subtropical fruit trees are mango, various species of citrus, avocado, papaya, and kiwi.

Scope and importance of fruit crops

Fruit farming is one of the important and age-old practices, practiced in India since ancient times. Cultivation of fruits plays an important role in the overall status of mankind and the nation. The standard of living of the people of a country is depending upon the production and per capita consumption of several fruits. Fruit growing also has more economic advantages.

Profitable fruit farming

High yield

First of all, you can earn a better return than many of the field crops from a well maintained and established fruit garden. From a unit area, you can realize more agronomic fruit crops.

High net profits

Though the initial cost of establishment of an orchard is height and it is compensated by higher productivity or due to the high value of produce.

Efficient utilization of resources

Agronomic fruit crops are seasonal and hence, farmers have to engage themselves in other occupations during slack seasons. However, fruit cultivation is being perennial. Then, it enables the grower to remain engaged throughout the year in farm operations and utilize full resources and assets like machinery, farm, land, water, etc.

Foreign trade potential

Many fresh fruits and processed products and spices have excellent export potential. Farmers can earn an excellent amount of foreign money from international trade.

Continuous cash flow

The harvesting process of most of the fruit is continuous. Due to the perishable nature, it wants to be marketed immediately after harvest. This gives a source of a continuous flow of inputs and other expenses of immediate nature as agronomic crops that are harvested one time.

Economic importance of fruit farming

High productivity – From a unit area of land more crop yield is realized from fruit crops than any of the agronomic crops. The average yields of Papaya, Banana and also Grapes are 10 to 15 times than that of agronomic crops.

High net profit – Through, the initial cost of establishment of an orchard is high, and it is compensated by higher net profit due to higher productivity or high value of produce.

Efficient utilization of resources – Growing of fruits being perennial, and enables the grower to remain engaged throughout the year in farm operations. And to utilize fully the resources and assets such as machinery, labor, and land water for production purposes throughout the year compared to agronomic crops.

Some of the factors affected by profitable fruit farming

Site selection for fruit farming

The site of a fruit-growing enterprise is important in determining its success as the varieties grown. Fruit variety and site together set a ceiling on the productivity and profit that can be realized under the best management. In most developed fruit regions microclimatic conditions such as climate at plant height, as influenced by slight differences in soil, soil cover, and elevation and soil conditions are the two components of a site that verify its desirability for a fruit-growing enterprise.

For highest productivity, most fruit crops must root extensively to a depth of three feet (one meter) or more. Heavy subsoil or other conditions causing imperfect internal drainage could effect in shallow, weak root systems that do not take water and nutrients efficiently from the soil.

Fertilizing

After you've planted your fruits, you'll need to give them enough water and food to grow and produce fruit. If your soil doesn't have sufficient nutrients for your plants, you'll want to provide fertilizer. You can use fertilizer in the form of compost or manure for example bat guano or chicken manure or you can purchase commercial fertilizer.

Growing fruit crops for Profit

Decide what fruits you are going to plant for your initial crop. Many factors play into this decision, but local climate, season and anticipated demand for the product must probably be given the most weight. Keep in mind that while some fruits such as strawberries can be grown quickly, developing fruit orchards is much more time- and capital-intensive enterprise, and it can be several years before producing your first cash crop.

Buy the necessary equipment and supplies to prepare your fruit garden. If you are growing organic be careful in choosing both your pest control and fertilizer means. Pest control for certain organic crops can be very challenging in some areas or climates. If space is not an issue, try to leave plenty of room between fruit plants and crop boundaries when planting for easy working conditions. Try to plant crops that will mature at different times so you will have some fresh fruits for sale most of the year.

Check on your garden regularly, daily if possible, and make sure you water and fertilize regularly as required for the fruits you are growing. Try to harvest on a staggered basis as much as possible, so you will have regular batches of produce for sale over the season of the different crops. Sell your fresh fruits as rapidly as possible.

Most profitable fruits in India

Profitable Aonla fruit farming in India

Aonla or amla is a very important minor fruit crop. The amla crop is quite hardy, prolific bearer and highly remunerative even without much care. The Aonla fruit is capsular (drupaceous) with a fleshy exocarp. In India, it is called as the Indian gooseberry.

India ranks first in the world in the area and production of this Aonla crop. Domestic consumers give the major market to Aonla.

Aonla is mostly cultivated in the states of Uttar Pradesh, Maharashtra, Rajasthan, Andhra Pradesh, Karnataka, Tamil Nadu, and Himachal Pradesh, etc. Aonla being a sub-tropical crop and prefers a dry subtropical climate. Heavy frost during the winter season is not appropriate for its cultivation. Aonla trees are hardy and stand well against drought. So, hardly any irrigation is practiced. However, the young plants need watering during summer, months at fortnightly intervals till they have fully established.

Slightly acidic to saline or sodic soil is suitable for Aonla cultivation. Aonla is not exacting in its soil requirement and grows well in sandy loam to clay soils in India. It has a great tolerance to salinity and sodicity and cultivated at a pH level of 6.0 to 8.0 very successful. Though, you will get a high-yield production in deep and fertile soils.

Profitable Banana fruit farming in India

Bananas are in popular demand in India, especially in the winter season. They are mostly grown in the states of Tamil Nadu, Gujarat, Maharashtra, Andhra Pradesh, and Karnataka. You will find banana selling everywhere, even in the summer. They're one of the most commonly consumed fruits in the country. The major feature of this plant is that it can be cultivated throughout the year. The high technology methods for banana farming can develop both yield and profit. Banana is a tropical plant and it requires a warm humid climate. However, you can grow banana in areas with temperatures ranging from 15 to 35°C.

The fertility of the soil is important for successful cultivation, as banana is a heavy feeder. Banana fruit is one of the few fruits, which has a restricted root zone. Thus, depth and drainage are the two most important considerations in selecting the soil for the banana. Bananas have importance to small-scale farmers in the developing countries of the tropics and sub-tropics. So, banana is popular for its low price and high nutritive value. One acre of plantain and banana has about 800 to 1,000 plants and the average production of Banana varies from 30 thousand fruit units for plantain and 100 thousand fruit units for banana per acre.

Profitable Mango fruit farming in India

Mangoes are the national fruit of India, and the mango tree is native to the Indian subcontinent. You will find mango plants all across India. The most popular season for growing mangoes in summer. You can start a profitable farming business by growing mangoes but be careful because there is a lot of competition already.

Mango is utilized at all stages of its development both in its immature and mature state. Mango fruit thrives well in tropical and sub-tropical climates. The ideal temperature for the mango crop ranges from 24° to 30° C along with high humidity. Temperature range below 10° C and above 42° C retards growth and adversely affects the flowering time of mango. Mango grows well on a wide variety of soils, for example lateritic, alluvial, sandy loam and sandy. Although it grows well in high to medium fertility soils. Mango trees need very less or no pruning. Though, the training of the plants in the initial stages is essential to give them the proper shape.

The expected mango produce is 10 tonnes per acre and the price will be around Rs.16,000/- per tonne. For one acre the revenue will be Rs.1.60 Lakhs and for 3 acres and it will be Rs.4.80 lakhs per year. The mango fruits will be borne from the 3rd year of planting the trees.

Profitable Guava fruit farming in India

Guava is a common tropical fruit and it is a nutritionally valuable remunerative crop. Commercial guava cultivation is a profitable agriculture business. Guava farming can be a very profitable agribusiness in India. Guava shows better growth under rain fed weather conditions; though, draining is equally essential. Depending on the climate conditions, during the dry season, irrigate the plants weekly once whereas in the winter, maintain 3 weeks. The drip irrigation system is ideal for better plant growth and void fungus infestation. Depending on the variety of guava, generally, cultivators find plants to carry fruits from 2 to 3 years of their aging, however, the vast majority attain their pick productivity point at the age of 7 to 10 years. However, the best soils are deep, friable and well-drained with a pH level between 5.5 and 7.5.

Profitable Papaya fruit farming in India

Papaya is a popular fruit grown in India. It's delicious and juicy and consumed most frequently in the summer. Papaya can also be cultivated in the winter. Papaya is cultivated most commonly in the states of Andhra Pradesh, Karnataka, Gujarat, Assam, Orissa, West Bengal, Kerala, Maharashtra, and Madhya Pradesh. It is also a large export to Gulf countries. Thus if you're looking for products to export from India to other countries, this is an option.

It is a popular fruit famous for its high nutritive and medicinal values. It comes early in bearing than any other fruit crop, produces fruits in less than a year and the production of papaya fruits is quite high per unit area.

Profitable Citrus fruit farming in India

Citrus is the highest valued fruit crop. Citrus orchards flourish well in light soils with excellent drainage properties. Deep soils with a pH level range of 5.5 to 7.5 are considered ideal. Though, they can also be grown in a pH range of 4.0 to 9.0. Citrus fruits grow best between temperature ranges of 13°C to 37°C. Citrus wants critical stage watering in the initial year. It further reduces fruit drop and then increases the fruit size. Diseases such as root rot and collar rot occur in flooded conditions.

Apple (Malus domestica)

Another fruit that can prove the right choice for commercial cultivation is apple. Due to the increasing demands, it makes for the most valuable crops, both nationally and internationally, and also has a high industrial requirement as well.

Grapes (Vitis vinifera)

One of the crucial fruits cultivated in India commercially, grapes always remains in high demand. It is also processed in industries for the production of raisins, jams, and liquor. Although meager amounts of the fruits are exported to other countries right now, the demand for Indian grapes is increasing day by day in international markets.

Coconut (Cocos nucifera)

Coconut is one of the important fruits cultivated in India and plays an integral part in the economy of the country as well. Every part of the fruit is used commercially in various industries like cosmetics, food, and medicines. The fruit also is an integral part of South Indian cuisine, which makes it an excellent choice for commercial cultivation.

Pomegranate (Punica granatum)

Another crop with high commercial value, many countries like the UAE, Bangladesh, and other European countries imports the fruit from India, which also happens to be the largest pomegranate producing country in the world, thereby expanding the scope for commercial cultivation of the fruit in India.

Watermelon (Citrullus lanatus)

The people of India have immense love for watermelons. Commercial cultivation of the fruit is very profitable, as the demand for the fruit is high in markets for its taste and health values. Apart from that, the fruit also has uses in various food and cosmetic industries.

Dragon fruit (Hylocereus undatus)

Dragonfruit or Pitaya is a variety of cactus, which is full of nutrients. Its commercial cultivation is gradually increasing with the growing demand in the country and outside. Cultivating dragon fruit is profitable in India, as it sells at high prices in Indian supermarkets.

Kiwifruit (Actinidia deliciosa)

Another fruit which can get you a good sum of profits is kiwifruit. The fruit has a high demand in Indian markets due to the growing health consciousness among people and it also sells for a good price, making it a smart choice for commercial cultivation!

5

HORTICULTURE OF BANANA AND APPLE

Banana Farming is gaining popularity day by day. Cultivation of bananas requires few simple steps but good management practices so as to earn a good profit out of it. There are many people, who would like to do the Banana Farming Business because, in recent times, it has become a common trend to have a farm either for the food crops, poultry & livestock, this is because of its high-profit yield.

Banana Farming is gaining popularity day by day. Cultivation of bananas requires few simple steps but good management practices so as to earn a good profit out of it. There are many people, who would like to do the Banana Farming Business because, in recent times, it has become a common trend to have a farm either for the food crops, poultry & livestock, this is because of its high-profit yield.

The banana industry is an important part of the global industrial agrobusiness. About 15% of the global banana production goes to export and international trade for consumption in Western countries. They are grown on banana plantations primarily in the Americas.

Banana is the fruit of a plant of the genus Musa (family Musaceae), which is cultivated primarily for food and secondarily for the production of fibre used in the textile industry are also cultivated for ornamental purposes. Almost all the modern edible parthenocarpic bananas come from the two wild species – Musa acuminataMusa balbisiana. The scientific names of bananas are Musa acuminata, Musa balbisiana or hybrids of Musa acuminata and balbisiana, depending on their genomic constitution.Bananas are vigorously growing, monocotyledonous herbaceous plants.The banana is not a tree but a high herb that can attain up to 15 meters of height.

The cultivars vary greatly in plant an d fruit size, plant morphology, fruit quality and disease and insect resistance. Most bananas have a sweet flavor when ripe; exceptions to this are cooking bananas and plantains. Plantains are hybrid bananas in which the male flowering axis is either degenerated, lacking, or possess relicts of male flowers. Plantains are always cooked before consumption and are higher in starch than bananas. The two groups of plantains, French and Horn, produce fewer fruit per plant than sweet bananas. The groups differ in whether the male parts of the inflorescence are present or absent.

Varieties

Dessert

Robusta, Dwarf Cavendish, Grand Naine, Rasthali, Vayal vazhai, Poovan, Nendran, Red Banana, Karpooravalli, Co.1, Matti, Sannachenkadali, Udayam and Neypoovan are popular varieties in banana. Cavendish groups are generally prefered in export market.

Culinary

Monthan, Vayal vazhai, Ash Monthan and Chakkia are cultivated for culinary purpose. Nendran is a dual purpose variety used for dessert and culinary.

Hill areas

The popular varieties of bananas suitable for hilly areas are Virupakshi, Sirumalai and Namarai. Red Banana, Manoranjitham (Santhana vazhai) and Ladan are also cultivated in hills.

Ecological impact

As is the case with all monocultures, the intensive agroindustrial methods for banana production have considerable environmental impact. Banana cultivation requires a large amount of pesticide (estimate: 35 lb/acre, 39 kg/ha) and can encourage ecosystem destruction through deforestation. Moreover, food miles and plastic packaging leave a large carbon footprint.

Cash crop

In 2012 the volume of global gross banana exports reached a record high of 16.5 million metric tons (3.6×1010 lb), 1.1 million tonnes (or 7.3 percent) above 2011 level. Bananas are the most popular fruit in the United States, with more consumed annually than apples and oranges combined. In spite of the multitude of banana species across the world, even only taking into account the cultivated ones, industrial production is dominated by the Cavendish banana.

Spread of plant disease

Tropical Race 4 is thought to be distributed globally by soil-contaminated equipment from the multinational plantation owners.

Companies

In 2013, five multinational fruit companies alone controlled 44% of the international banana trade:

Company	Country Market share in 2013 [%]
Chiquita	
U.S.	13
Fyffes	
Ireland	6
Dole Food Company	
U.S.	11
Fresh Del Monte Produce	
U.S.	12
Noboa	
Ecuador	2

The market share of the above players decreased from 70% in 2002 to about 44% in 2013. This decline in market power has been attributed to a couple of reasons. In the past, multinational companies owned a large number of plantations in Central and South America and other banana-producing regions. Since the 1980s they have divested a large share of their own production, replacing it with greater purchases from independent producers. For example, Chiquita has decreased the number of its plantations in Central America. Fyffes used to own plantations in Jamaica, Belize and the Windward Islands, but withdrew from production and switched to purchasing its bananas through contracts with producers. The disengagement from production was partly caused by legal and economic problems at the plantation level, but also reflects the change in market power along the banana value chain. Along the global banana supply chain major supermarket chains in the US and EU have gained market power over the big producers in the 21st century as they dominate the retail market and increasingly purchase from smaller wholesalers or directly from growers.

Production and export

In 2016, world production of bananas and plantains was 148 million tonnes, led by India and China with a combined total (only for bananas) of 28% of global production (table). Other major producers were the Philippines, Ecuador, Indonesia, and Brazil, together accounting for 20% of the world total of bananas and plantains (table).

As reported for 2013, total world exports were 20 million tonnes of bananas and 859,000 tonnes of plantains. Ecuador and the Philippines were the leading exporters with 5.4 and 3.3 million tonnes, respectively, and the Dominican Republic was the leading exporter of plantains with 210,350 tonnes.

Controversy

The global banana industry has been involved with multiple human rights violations since its inception in the late 19th century. These include slavery and forced child labor among others and as of 2021, the Rainforest Alliance has identified Côte d'Ivoire as the country with the highest risk of forced labor, followed by Cameroon, Ghana, Guatemala and Ecuador among other banana exporting countries.

By following basic guidelines you can earn a really big income through banana cultivation. Below are the tips that will guide you in Starting your own Banana Farm at any location in the whole world:

Have Sufficient Money

- Money takes the lead in starting any kind of business in the whole wide world. It does not matter if you want to learn the business tips or start a business, you will definitely have to invest a certain amount to acquire the training or to start the business itself.
- In the farming business, whether it is small-scale farming or large-scale farming, you must have a certain amount with yourself to tackle any kind of financial need, which may arrive at any point of time.
- One should plan very well before starting a business, you can choose to start with a small & slowly and gradually advance it later when you made enough cash from the business.

Get Appropriate Farmland

- The banana plantation is one of the kinds of commercial agricultural farming, which requires an appropriate site facility under the tropical climatic conditions to give the best of its productivity.
- For banana cultivation, you can buy land, which is cheaper in rural or suburban areas, or you can or look for farmers, who can share the land with you. Keep in mind that the land should have the right kind of soil for banana cultivation.
- Be mindful while selecting fertile land for the banana plantation, because the fertility of land will lead you to determine the kind of yield, which you will get from your banana farm.

Soil Check

- Below are the tips, which you will need to follow while starting a banana plantation. One cannot discover an empty land and start a banana plantation, without knowing the type of soil that a banana plantation requires to do well.
- After you have acquired land, check if the soil meets up with the type of soil that you require for the banana plant to survive and provide a better yield.
- Banana can grow from the poor to the richest of land, but the success depends on the type of soil that banana farming requires a good fertility land along with good moisture content & drainage.

- The banana cultivation requires a rich loamy soil accompanied with pH values ranging from 6 – 7.5. The soil should have a good quality of adequate nitrogen, potash, phosphorus & some organic contents, not too much alkaline or acidic.

Note: The land should undergo a soil test before using it for commercial banana farming. Saline, Calcareous & Sandy soil are not good for a Banana plantation.

Banana Sucker Treatment

- The sucker is first of all planted in a nursery, and then transplants this into the already prepared moist soil & makes sure to space out properly. The banana sucker is a small shoot, comes from the base of banana plants. Every species of the banana have suckers, where the plantation emerges. It should be well-trimmed & treated before putting in the soil.
- It is advisable to treat the sucker as it helps to get rid of pests & disease that attacks the sucker at the zone. The sucker can be treated by the method of red acalypha leaf extract or boiling water treatment.

Planting Season of Banana

- Banana planting is the best during the late spring & early summer season around May-June or in the month of autumn September-October.
- Banana Farming suggests when temperature increases it will accelerate disease development in bananas. The distribution of rainfall is said to have a larger impact on the harshness of the banana disease.

Cost & Profit in banana cultivation

In banana cultivation, about 3000 plants are planted in one hectare i.e. up to Rs. 45000-60000, you will have to spend only on plants. At the same time, around Rs. 2.5-3 lakh per hectare is also spent in the care & management of plants throughout the year.

Cultivation of bananas in one hectare will cost about Rs. 3-4 lakh (for entire 12-14 months). Moreover, up to 25-40 kg of bananas come out per plant. In this way, about 100 tonnes of bananas are produced from one hectare.

These bananas sell between Rs 10-15 per kg. Assuming an average price of Rs 12, you will earn Rs 12 lakh. On the other hand, if we remove the cost, you get a profit of Rs 8 lakh.

Banana farming is considered extremely beneficial not because it serves as a source of food. It also creates self-employment for farmers to become banana exporters & renowned suppliers to the beverage companies.

Planting

Preparation of land:

- Plough the land thoroughly atleast for 3-4 times and add about 10 tonnes of well rotten FYM or Compost during last plough and mix it well or add 10-15 kg FYM/ Compost per pit of 60x60x60 cm dimension.

Selection of Suckers:

- Select 'Sword Suckers' with broad corm with narrow sword like leaves, from viral, fungal and bacterial disease free mother plants.
- The suckers should be 3-5 months old, uniform in size, weighing 1-1.5 kg for Nendran, Rasthali, Ney Poovan and Poovan Banana varieties.
- For long duration varieties like Karpuravalli and Red Banana, slightly big suckers weighing 1.5-2.0 kg should be used.
- For planting of 'Tissue Culture' plants, the secondary hardened plant should be about 30 cm tall, 5 cm girth with atleast five fully opened healthy leaves and true to type.

Sucker Treatment and planting:

- The selected suckers should be 'pared' by trimming of all the roots along with surface layers superficially to remove any rotten portion of the corm.
- Dip the pared suckers in 0.2% Carbendazim (2g/litre of water) solution for about 15 –20 minutes as a prophylactic measure against Fusarium wilt disease.
- Keep the treated suckers in shade overnight before planting. Plant the suckers in the center of the pit and press the soil around the suckers firmly.
- Apply 40 g of Carbofuron granules per pit to protect the plants against nematode attack and irrigate the field thoroughly.
- In case of tissue culture plants, one week before planting apply 10 g Carbofuron and 1.0 % bleaching powder or 0.2 % Emissan in 100 ml water as drench into the polythene bags to protect against nematode infestation and bacterial rot (Erwinia Rot) disease respectively.

Calendar of Operations

During First month

- The soil around the plants should be pressed firmly for better and quick establishment of the plants.
- Wherever necessary, 'Gap Filling' should be done to replace the un- sprouted as well as rotten suckers. Seeds of green manure crops viz., cowpea or sunnhemp be sown.

- For additional income and also for effective land use efficiency, short duration crops such as onion, green gram, black gram, beans, radish, greens, marigold and short duration vegetables can be grown as intercrop.
- Tomato, chilly and cucurbits should not be grown as intercrop since these crops harbour nematodes and aphids, which act as vector of virus spread.

Second Month

- Green manures viz., cowpea or sunnhemp should be ploughed back in to the soil at flowering stage or about 40 days after sowing.
- Slight digging and earthing up to keep the weeds under control.
- For Fusarium wilt susceptible varieties like Rasthali, Karpuravalli, Ney Poovan, Monthan and Pachanadan, drench the soil around the plant with 0.2% Carbendazim as a prophylactic measure, or
- Apply 30g Trichoderma viride or Pseudomonas flourescense along with FYM/ compost 1 kg in the soil around the plant as a prophylactic measure for the control of wilt disease.

Third Month

- Application of 40g of Carbofuron to control nematodes.
- Digging and weeding.
- Application of first dose of fertilizers @ 100:300:100 g Urea, Super Phosphate and MOP per plant in basins made about 30 cm away from the plant.

Fourth Month

- Application of Azospirillum and phosphobacteria @ 30 g and Trichoderma viride @ 30g along with 5-10 kg FYM plant-1.
- There should a gap of minimum 2-3 weeks between the application of chemical fertilizers and biofertilizers.
- Periodical removal of side suckers by cutting them above the ground level and pouring 2 ml kerosene at the central core of the sucker.
- If any virus affected plants are noticed in the field, remove and destroy it immediately and spray with any systemic insecticide to kill the insect vectors which spread the virus.

Fifth Month

- Application of second dose of fertilizers @ 150:150 g Urea and MOP+ 300g neemcake per plant in the basins made about 45 cm away from the plant.
- Removal of dried leaves.

- Digging and weeding.
- To cater the micronutrient need of the plant and to correct their deficiency, apply 50g agricultural lime and 25g magnesium sulphate per plant.
- For Fusarium wilt susceptible varieties like Rasthali, Karpuravalli, Ney Poovan, Monthan and Pachanadan drench the soil around the plant with 0.2% Carbendazim as a prophylactic measure.
- To prevent the egg laying and further attack of stem weevil, spray 'Neemosol' @12.5ml/litre or Chlorpyriphos @ 2.5ml/litre on the stem especially in Nendran, Red Banana, Karpuravalli and Monthan varieties.
- To monitor the corm and stem weevil, 2 ft long longitudinal stem trap @40 traps/acre can be placed at different places. The collected weevils are to be killed using kerosene.
- Keep the Banana fields as well as surrounding areas weed free and spray systemic insecticides to control the insect vectors.

Sixth Month

- Digging and earthing up of soil around the plant.
- Removal of the dried and diseased leaves and spraying of 0.1% Propiconazol (TILT) by thoroughly covering both the surfaces adding wetting agent with the spray fluid especially during winter and cool months for control of Sigatoka leaf spot diseases.
- Yellowing of leaves which is a symptom of iron deficiency, spray 0.5% ferrous sulphate + 1.0% urea added with wetting agent on the leaves especially in high pH >8.5 and Calcareous soils.
- To correct the deficiency of zinc, spray 0.5% zinc sulphate solution along with wetting agent.
- Foliar application of 0.5 Borax is recommended to correct the deficiency.
- Apply 30g Trichoderma viride or Pseudomonas flourescense in the soil around the plant as a prophylactic measure to control the wilt disease.
- For controlling the stem weevil attack, using 'Banana Injector', inject 2ml of Monocrotophos (150 ml Monocrotophos mixed in 350 ml of water) at 2 and 4 feet height on opposite direction.

Seventh Month

- Application of third dose of fertilizers @ 150:150 g Urea and MOP per plant in the basins made about 60 cm away from the plant.
- Removal of the dried and diseased leaves and spraying of 0.1% Carbendazim or Calixin by thoroughly covering both the surfaces along with wetting agent.

- Periodical removal of side suckers by cutting them above the ground level, scoop the core and pour 2 ml kerosene in the core.
- Injection of 2ml of Monocrotophos using 'Banana Injector' at 2 and 4 feet height for the control of stem weevil.

Eighth Month

- After flowering, only one healthy side sucker should be allowed for first ratoon and the remaining suckers should be killed using kerosene or uprooted.
- Spraying of 0.1% Indofil by thoroughly covering both the surfaces.
- After the emergence of the last hand, the male bud has to be removed leaving about 15 cm stalk from the last hand.
- To prevent 'cigar end rot' disease, remove the pistil and perianth carefully from the fully emerged fingers and spray the bunch with Indofil M-45 @ 2.5 ml/litre.
- Spray 2% Potassium Sulphate (20g/litre of water) solution with surfactant by thoroughly drenching the bunch and cover the bunch with 100 gauge thick white or blue polythene sleeves having 6% ventilation.

Ninth Month

- Thirty days after the first spray, give a second spray of 2% Potassium Sulphate (20g/litre of water) solution with surfactant by thoroughly drenching the bunch.
- Provide casuarina pole or bamboo support to the plants for tall and heavy bearing bunches.

Cultivation of Tissue Culture Banana

The propagation of a plant by using a plant part or single cell or group cell in a test tube under very controlled and hygienic conditions is called "Tissue Culture". In India, tissue culture banana is grown in large scale. Particularly, Grandnaine variety is a preferred variety by farmers.

Land Preparation

Prior to planting banana, grow the green manuring crop like daincha, cowpea etc. and burry it in the soil. The land can be ploughed 2-4 times and leveled. Use ratovator or harrow to break the clod and bring the soil to a fine tilt. During soil preparation basal dose of FYM is added and thoroughly mixed into the soil.

A pit size of 45cm x 45cm x 45cm is normally required. The pits are to be refilled with topsoil mixed with 10 kg of FYM (well decomposed), 250 gm of Neem cake and 20 gm of conbofuron. Prepared pits are left to solar radiation helps in killing the harmful insects, is effective against soil borne diseases and aids aeration. In saline alkali soil where PH is above 8 Pit mixture is to be modified to incorporate organic matter.

Addition of organic matter helps in reducing salinity while addition of purlite improves, porosity and aeration. Alternative to planting in pits is planting in furrows. Depnding on soil strata one can choose appropriate method as well as spacing and depth at which plant is required to be planted.

Planting material

Sword suckers weighing approximately 500-1000 gm are commonly used as propagating material. Suckers generally may be infected with some pathogens and nematodes. Similarly due to the variation in age and size of sucker the crop is not uniform, harvesting is prolonged and management becomes difficult.

Therefore, in-vitro clonal propagation i.e. Tissue culture plants are recommended for planting. They are healthy, disease free, uniform and authentic. Properly hardened secondary seedlings are only recommended for planting

Advantages of Tissue Culture planting material

- True to the type of mother plant under well management.
- Pest and disease free seedlings.
- Uniform growth, increases yield.
- Early maturity of crop - maximum land use is possible in low land holding country like India.
- Round the year planting possible as seedlings are made available throughout the year.
- Two s uccessive ratoons are possible in a short duration which minimizes cost of cultivation.
- No staggered harvesting.
- 95% - 98% plants bear bunches.
- New varieties can be introduced and multiplied in a short duration.

Planting time

Planting of tissue culture Banana can be done throughout the year except when the temperature is too low or too high. Facility of drip irrigation system is important. There are two important seasons in Maharashtra, India;

1. Mrig Baug (Kharif) Month of planting June - July.
2. Kande Baug (Rabi) Month of planting October - November.

Crop Geometry

Traditionally banana growers plant the crop at 1.5m x 1.5m with high density, however plant growth and yields are poor because of competition for sunlight. Various trials are conducted at Jain Irrigation System R&D farm with Grandnaine as cultivar. And then suitable spacing of 1.82m x 1.52m is being recommended, it accommodates 1452 plants per acre (3630 plants per hectare) keeping row direction North-South with wide spacing 1.82m between the rows. The region like north India, coastal belt and where humidity is very high and temp falls down upto 5-7°C, the planting distance should not be less than 2.1m x 1.5m.

Planting method

Polybags is separated from the plant without disturbing the root ball of the plant and then plants are planted in the pits keeping the pseudo-stem 2cm below the ground level. Soil around the plant is gently pressed. Deep planting should be avoided.

Water management

Banana, a water loving plant, requires a large quantity of water for maximum productivity. But Banana roots are poor withdrawal of water. Therefore under Indian condition banana production should be supported by an efficient irrigation system like drip irrigation.

Water requirement of banana has been worked out to be 2000mm per Annum. Application of drip irrigation and mulching technology has reported improved water use efficiency. There is saving of 56% of water and increasing yield by 23-32% under drip.

Fertigation

Banana requires high amount of nutrients, which are often supplied only in part by the soil. Nutrient requirement has been worked out on all India basis is to be 20 kg FYM, 200gm N; 60-70gm P; 300gm K/plant. Banana requires heavy nutrition. Banana crop requires 7-8 Kg N, 0.7- 1.5 Kg P and 17-20 Kg K per metric tonne yield. Banana responds well to application of nutrients. Traditionally farmers use more of urea and less of phosphorous and potash.

In order to avoid loss of nutrients from conventional fertilizers i.e. loss of N through leaching, volatilization, evaporation and loss of P and K by fixation in the soil, application of water soluble or liquid fertilizers through drip irrigation (fertigation) is encouraged. A 25-30% increase in yield is observed using fertigation. Moreover, it saves labour and time and the distribution of nutrients is uniform.

Weeding

Spraying of Glyphosate (Round up) before planting at the rate of 2 lit/ha is carried out to keep the plantation weed free. One or two manual weedings are necessary.

Micronutrient Foliar Spray

Combined foliar application of ZnSo4 (0.5%), FcSo4 (0.2%), CuSo4 (0.2%) and H3Bo3 (0.1%) can be adopted to improve morphological, physiological and yield attributes of banana. The micronutrient spray solution is prepared by dissolving the following in 100 lit. of water.

Special Operations

There are operation specific to banana crop which influence productivity and quality.

Desuckering

Removal of unwanted suckers is a critical operation in banana for reducing internal competition with mother plant.

Desuckering should be done regularly until shooting. However in areas where ratoon is also taken for the second crop, a follower is allowed after inflorescence has appeared and this should be managed that planting space is not disturbed. Follower should be opposite to the inflorescence. It should not be far apart from the main plant.

Deflowering

It consists of removal of the withered style and perianth. This is generally not practiced. Therefore, they remain attached to the fruit bunch & then removed after harvesting which is damaging to the fruits. It is therefore suggested that you remove them just after flowering.

Pruningof leaves

Rubbing leaves damages the fruit, therefore, such leaves should also be pruned during regular check. Older leaves and infected leaves also be pruned as required. Green leaves should not be removed.

Earthing up

Keep the soil loose by harrowing from time to time. Earthing up should be done at 3-4 months after planting i.e. raising the soil level around the base of the plant by 10-12". It is better to prepare a raised bed and keep the drip line on bed 2-3" away from the plant. It also helps to protect plants from wind damage and production losses to some extent..

Removal of male buds

(Denavelling) Removal of male buds helps fruit development and increases bunch weight. Male buds are removed from the last 1-2 small hands with a clean cut keeping a single finger in the last hand.

Bunch Spray

Spray of monocrotophos (0.2%) after emergence of all hands takes care of the thrips. Thrips attack discolors the fruit skin and makes it unattractive.

Bunch Covering

Covering bunch using dried leaves of the plant is economical and prevents the bunch from direct exposure to sunlight. Bunch cover enhances quality of fruit. But in rainy season this practice should be avoided.

Sleeving of bunch is done to protect fruits against dust, spray residue, insect and birds. For this blue plastic sleeves are preferred. This also increases temperature around developing bunch and helps in early maturity.

Dehandling of false hands of bunch

In a bunch there are some incomplete hands which are not fit for quality produce. These hands should be removed soon after bloom. This helps in improving the weight of other hands. Sometimes the hand just above the false hand is also removed.

Propping

Due to heavy weight of bunch the plant goes out of balance and the bearing plant may lodge and production and quality are adversely affected. Therefore they should be propped with the help of two bamboos forming a triangle by placing them against the stems on the leaning side. This also helps in uniform development of bunch.

Harvesting

Banana should be harvested at the physiological maturity stage for better post harvest quality. The fruit is climacteric and can reach consumption stage after ripening operation

Maturity indices

These are established on the basis of fruit shape, angularity, grade or diameter of the median figure of the second hand, starch content and number of days that have elapsed after flowering. Market preferences can also affect the decision for harvesting a slight or full mature fruit.

Removal of bunch

Bunch should be harvested when figures of second hand from top are 3/4 rounded with the help of sharp sickle 30cm above the first hand. Harvest may be delayed upto 100-110 days after opening of the first hand. Harvested bunch should generally be collected in well padded tray or basket and brought to the collection site. Bunches should be kept out of light after harvest, since this hastens ripening and softening.

For local consumption, hands are often left on stalks and sold to retailers.

For export, hands are cut into units of 4-16 fingers, graded for both length and girth, and carefully placed in polylined boxes to hold different weight depending on export requirements.

Post harvest operations

At collection site injured and over mature fruits are discarded and for local market bunches should be delivered through lorries or wagons. However, for more sophisticated and export market where the quality is predominant, bunches should be dehanded, fruits are cleared in running water or dilute sodium hypochlorite solution to remove the latex and treated with thiobendasole; air dried and graded on the basis of size of fingers as already stated, packed in ventilated CFB boxes of 14.5 kg capacity or as per requirement with polythene lining and pre-cooled at 13-15°C temperature and at 80-90% RH.

Such material should than be sent under cool chain at 13°C for marketing

Yield

The planted crop gets ready for harvest within 11-12 months of planting. First ratoon crop would be ready by 8-10 month from the harvesting of the main crop and second ratoon by 8-9 months after the second crop.

Thus over a period to 28-30 months, it is possible to harvest three crops i.e. one main crop and two ratoon crops. Under drip irrigation combined with Fertigation yield of Banana as high as 100 T/ha can be obtained with the help of tissue culture technique, even similar yield in the ratoon crops can be achieved if the crop is managed well.

APPLE HORTICULTURE

An apple is an edible fruit produced by an apple tree (Malus domestica). Apple trees are cultivated worldwide and are the most widely grown species in the genus Malus. The tree originated in Central Asia, where its wild ancestor, Malus sieversii, is still found today. Apples have been grown for thousands of years in Asia and Europe and were brought to North America by European colonists. Apples have religious and mythological significance in many cultures, including Norse, Greek, and European Christian tradition.

Apples grown from seed tend to be very different from those of their parents, and the resultant fruit frequently lacks desired characteristics. Generally, apple cultivars are propagated by clonal grafting onto rootstocks. Apple trees grown without rootstocks tend to be larger and much slower to fruit after planting. Rootstocks are used to control the speed of growth and the size of the resulting tree, allowing for easier harvesting.

There are more than 7,500 known cultivars of apples. Different cultivars are bred for various tastes and uses, including cooking, eating raw, and cider production. Trees and fruit are prone to a number of fungal, bacterial, and pest problems, which can be controlled by a number of organic and non-organic means. In 2010, the fruit's genome was sequenced as part of research on disease control and selective breeding in apple production.

Worldwide production of apples in 2018 was 86 million tonnes, with China accounting for nearly half of the total.

Description

The apple is a deciduous tree, generally standing 2 to 4.5 m (6 to 15 ft) tall in cultivation and up to 9 m (30 ft) in the wild. When cultivated, the size, shape and branch density are determined by rootstock selection and trimming method. The leaves are alternately arranged dark green-colored simple ovals with serrated margins and slightly downy undersides.

Blossoms are produced in spring simultaneously with the budding of the leaves and are produced on spurs and some long shoots. The 3 to 4 cm (1 to 1+1⁄2 in) flowers are white with a pink tinge that gradually fades, five petaled, with an inflorescence consisting of a cyme with 4–6 flowers. The central flower of the inflorescence is called the "king bloom"; it opens first and can develop a larger fruit.

The fruit is a pome that matures in late summer or autumn, and cultivars exist in a wide range of sizes. Commercial growers aim to produce an apple that is 7 to 8.5 cm (2+3⁄4 to 3+1⁄4 in) in diameter, due to market preference. Some consumers, especially those in Japan, prefer a larger apple, while apples below 5.5 cm (2+1⁄4 in) are generally used for making juice and have little fresh market value. The skin of ripe apples is generally red, yellow, green, pink, or russetted, though many bi- or tri-colored cultivars may be found. The skin may also be wholly or partly russeted i.e. rough and brown. The skin is covered in a protective layer of epicuticular wax. The exocarp (flesh) is generally pale yellowish-white, though pink, yellow or green exocarps also occur.

Wild ancestors

The original wild ancestor of Malus domestica was Malus sieversii, found growing wild in the mountains of Central Asia in southern Kazakhstan, Kyrgyzstan, Tajikistan, and northwestern China. Cultivation of the species, most likely beginning on the forested flanks of the Tian Shan mountains, progressed over a long period of time and permitted secondary introgression of genes from other species into the open-pollinated seeds. Significant exchange with Malus sylvestris, the crabapple, resulted in current populations of apples being more related to crabapples than to the more morphologically similar progenitor Malus sieversii. In strains without recent admixture the contribution of the latter predominates.

Genome

Apples are diploid (though triploid cultivars are not uncommon), has 17 chromosomes and an estimated genome size of approximately 650 Mb. Several whole genome sequences have been made available, the first one in 2010 was based on the diploid cultivar 'Golden Delicious'. However, this first whole genome sequence turned out to contain several errors in part owing to the high degree of heterozygosity in diploid apples which, in combination with an ancient genome duplication, complicated the assembly. Recently, double- and trihaploid individuals have been sequenced, yielding whole genome sequences of higher quality.

The first whole genome assembly was estimated to contain around 57,000 genes, though the more recent genome sequences support more moderate estimates between 42,000 and 44,700 protein-coding genes. Among other things, the availability of whole genome sequences has provided evidence that the wild ancestor of the cultivated apple most likely is Malus sieversii. Re-sequencing of multiple accessions has supported this, while also suggesting extensive introgression from Malus sylvestris following domestication.

Significance in European cultures and societies

Germanic paganism

In Norse mythology, the goddess Iðunn is portrayed in the Prose Edda (written in the 13th century by Snorri Sturluson) as providing apples to the gods that give them eternal youthfulness. The English scholar H. R. Ellis Davidson links apples to religious practices in Germanic paganism, from which Norse paganism developed. She points out that buckets of apples were found in the Oseberg ship burial site in Norway, that fruit and nuts (Iðunn having been described as being transformed into a nut in Skáldskaparmál) have been found in the early graves of the Germanic peoples in England and elsewhere on the continent of Europe, which may have had a symbolic meaning, and that nuts are still a recognized symbol of fertility in southwest England.

Davidson notes a connection between apples and the Vanir, a tribe of gods associated with fertility in Norse mythology, citing an instance of eleven "golden apples" being given to woo the beautiful Gerðr by Skírnir, who was acting as messenger for the major Vanir god Freyr in stanzas 19 and 20 of Skírnismál. Davidson also notes a further connection between fertility and apples in Norse mythology in chapter 2 of the Völsunga saga: when the major goddess Frigg sends King Rerir an apple after he prays to Odin for a child, Frigg's messenger (in the guise of a crow) drops the apple in his lap as he sits atop a mound. Rerir's wife's consumption of the apple results in a six-year pregnancy and the birth (by Caesarean section) of their son—the hero Völsung.

Further, Davidson points out the "strange" phrase "Apples of Hel" used in an 11th-century poem by the skald Thorbiorn Brúnarson. She states this may imply that the apple was thought of by Brúnarson as the food of the dead. Further, Davidson notes that the potentially Germanic goddess Nehalennia is sometimes depicted with apples and that parallels exist in early Irish stories. Davidson asserts that while cultivation of the apple in Northern Europe extends back to at least the time of the Roman Empire and came to Europe from the Near East, the native varieties of apple trees growing in Northern Europe are small and bitter. Davidson concludes that in the figure of Iðunn "we must have a dim reflection of an old symbol: that of the guardian goddess of the life-giving fruit of the other world."

Greek mythology

Apples appear in many religious traditions, often as a mystical or forbidden fruit. One of the problems identifying apples in religion, mythology and folktales is that the word "apple" was used as a generic term for all (foreign) fruit, other than berries, including nuts, as late as the 17th century. For instance, in Greek mythology, the Greek hero Heracles, as a part of his Twelve Labours, was required to travel to the Garden of the Hesperides and pick the golden apples off the Tree of Life growing at its center.

The Greek goddess of discord, Eris, became disgruntled after she was excluded from the wedding of Peleus and Thetis. In retaliation, she tossed a golden apple inscribed Καλλίστη (Kalliste, sometimes transliterated Kallisti, "For the most beautiful one"), into the wedding party. Three goddesses claimed the apple: Hera, Athena, and Aphrodite. Paris of Troy was appointed to select the recipient. After being bribed by both Hera and Athena, Aphrodite tempted him with the most beautiful woman in the world, Helen of Sparta. He awarded the apple to Aphrodite, thus indirectly causing the Trojan War.

The apple was thus considered, in ancient Greece, sacred to Aphrodite. To throw an apple at someone was to symbolically declare one's love; and similarly, to catch it was to symbolically show one's acceptance of that love. An epigram claiming authorship by Plato states:

I throw the apple at you, and if you are willing to love me, take it and share your girlhood with me; but if your thoughts are what I pray they are not, even then take it, and consider how short-lived is beauty.

— Plato, Epigram VII

Atalanta, also of Greek mythology, raced all her suitors in an attempt to avoid marriage. She outran all but Hippomenes (also known as Melanion, a name possibly derived from melon, the Greek word for both "apple" and fruit in general), who defeated her by cunning, not speed. Hippomenes knew that he could not win in a fair race, so he used three golden apples (gifts of Aphrodite, the goddess of love) to distract Atalanta. It took all three apples and all of his speed, but Hippomenes was finally successful, winning the race and Atalanta's hand.

Christian art

Though the forbidden fruit of Eden in the Book of Genesis is not identified, popular Christian tradition has held that it was an apple that Eve coaxed Adam to share with her. The origin of the popular identification with a fruit unknown in the Middle East in biblical times is found in confusion between the Latin words mālum (an apple) and mălum (an evil), each of which is normally written malum. The tree of the forbidden fruit is called "the tree of the knowledge of good and evil" in Genesis 2:17, and the Latin for "good and evil" is bonum et malum.

Renaissance painters may also have been influenced by the story of the golden apples in the Garden of Hesperides. As a result, in the story of Adam and Eve, the apple became a symbol for knowledge, immortality, temptation, the fall of man into sin, and sin itself. The larynx in the human throat has been called the "Adam's apple" because of a notion that it was caused by the forbidden fruit remaining in the throat of Adam. The apple as symbol of sexual seduction has been used to imply human sexuality, possibly in an ironic vein.

Proverb

The proverb, "An apple a day keeps the doctor away", addressing the supposed health benefits of the fruit, has been traced to 19th-century Wales, where the original phrase was "Eat an apple on going to bed, and you'll keep the doctor from earning his bread". In the 19th century and early 20th, the phrase evolved to "an apple a day, no doctor to pay" and "an apple a day sends the doctor away"; the phrasing now commonly used was first recorded in 1922. Despite the proverb, there is no evidence that eating an apple daily has any significant health effects.

Cultivars

There are more than 7,500 known cultivars (cultivated varieties) of apples. Cultivars vary in their yield and the ultimate size of the tree, even when grown on the same rootstock. Different cultivars are available for temperate and subtropical climates. The UK's National Fruit Collection, which is the responsibility of the Department of Environment, Food, and Rural Affairs, includes a collection of over 2,000 cultivars of apple tree in Kent. The University of Reading, which is responsible for developing the UK national collection database, provides access to search the national collection. The University of Reading's work is part of the European Cooperative Programme for Plant Genetic Resources of which there are 38 countries participating in the Malus/Pyrus work group.

The UK's national fruit collection database contains much information on the characteristics and origin of many apples, including alternative names for what is essentially the same "genetic" apple cultivar. Most of these cultivars are bred for eating fresh (dessert apples), though some are cultivated specifically for cooking (cooking apples) or producing cider. Cider apples are typically too tart and astringent to eat fresh, but they give the beverage a rich flavor that dessert apples cannot.

Commercially popular apple cultivars are soft but crisp. Other desirable qualities in modern commercial apple breeding are a colorful skin, absence of russeting, ease of shipping, lengthy storage ability, high yields, disease resistance, common apple shape, and developed flavor. Modern apples are generally sweeter than older cultivars, as popular tastes in apples have varied over time. Most North Americans and Europeans favor sweet, subacid apples, but tart apples have a strong minority following. Extremely sweet apples with barely any acid flavor are popular in Asia, especially the Indian subcontinent.

Old cultivars are often oddly shaped, russeted, and grow in a variety of textures and colors. Some find them to have better flavor than modern cultivars, but they may have other problems that make them commercially unviable—low yield, disease susceptibility, poor tolerance for storage or transport, or just being the "wrong" size. A few old cultivars are still produced on a large scale, but many have been preserved by home gardeners and farmers that sell directly to local markets. Many unusual and locally important cultivars with their own unique taste and appearance exist; apple conservation campaigns have sprung up around the world to preserve such local cultivars from extinction. In the United Kingdom, old cultivars such as 'Cox's Orange Pippin' and 'Egremont Russet' are still commercially important even though by modern standards they are low yielding and susceptible to disease.

Cultivation

Breeding

Many apples grow readily from seeds. However, more than with most perennial fruits, apples must be propagated asexually to obtain the sweetness and other desirable characteristics of the parent. This is because seedling apples are an example of "extreme heterozygotes", in that rather than inheriting genes from their parents to create a new apple with parental characteristics, they are instead significantly different from their parents, perhaps to compete with the many pests. Triploid cultivars have an additional reproductive barrier in that three sets of chromosomes cannot be divided evenly during meiosis, yielding unequal segregation of the chromosomes (aneuploids). Even in the case when a triploid plant can produce a seed (apples are an example), it occurs infrequently, and seedlings rarely survive.

Because apples are not true breeders when planted as seeds, although cuttings can take root and breed true, and may live for a century, grafting is usually used. The rootstock used for the bottom of the graft can be selected to produce trees of a large variety of sizes, as well as changing the winter hardiness, insect and disease resistance, and soil preference of the resulting tree. Dwarf rootstocks can be used to produce very small trees (less than 3.0 m or 10 ft high at maturity), which bear fruit many years earlier in their life cycle than full size trees, and are easier to harvest.

Dwarf rootstocks for apple trees can be traced as far back as 300 BCE, to the area of Persia and Asia Minor. Alexander the Great sent samples of dwarf apple trees to Aristotle's Lyceum. Dwarf rootstocks became common by the 15th century and later went through several cycles of popularity and decline throughout the world. The majority of the rootstocks used today to control size in apples were developed in England in the early 1900s. The East Malling Research Station conducted extensive research into rootstocks, and today their rootstocks are given an "M" prefix to designate their origin. Rootstocks marked with an "MM" prefix are Malling-series cultivars later crossed with trees of 'Northern Spy' in Merton, England.

Most new apple cultivars originate as seedlings, which either arise by chance or are bred by deliberately crossing cultivars with promising characteristics. The words "seedling", "pippin", and "kernel" in the name of an apple cultivar suggest that it originated as a seedling. Apples can also form bud sports (mutations on a single branch). Some bud sports turn out to be improved strains of the parent cultivar. Some differ sufficiently from the parent tree to be considered new cultivars.

Since the 1930s, the Excelsior Experiment Station at the University of Minnesota has introduced a steady progression of important apples that are widely grown, both commercially and by local orchardists, throughout Minnesota and Wisconsin. Its most important contributions have included 'Haralson' (which is the most widely cultivated apple in Minnesota), 'Wealthy', 'Honeygold', and 'Honeycrisp'.

Apples have been acclimatized in Ecuador at very high altitudes, where they can often, with the needed factors, provide crops twice per year because of constant temperate conditions year-round.

Pollination

Apples are self-incompatible; they must cross-pollinate to develop fruit. During the flowering each season, apple growers often utilize pollinators to carry pollen. Honey bees are most commonly used. Orchard mason bees are also used as supplemental pollinators in commercial orchards. Bumblebee queens are sometimes present in orchards, but not usually in sufficient number to be significant pollinators.

There are four to seven pollination groups in apples, depending on climate:

- **Group A** – Early flowering, 1 to 3 May in England ('Gravenstein', 'Red Astrachan')
- **Group B** – 4 to 7 May ('Idared', 'McIntosh')
- **Group C** – Mid-season flowering, 8 to 11 May ('Granny Smith', 'Cox's Orange Pippin')

- **Group D** – Mid/late season flowering, 12 to 15 May ('Golden Delicious', 'Calville blanc d'hiver')
- **Group E** – Late flowering, 16 to 18 May ('Braeburn', 'Reinette d'Orléans')
- **Group F** – 19 to 23 May ('Suntan')
- **Group H** – 24 to 28 May ('Court-Pendu Gris' – also called Court-Pendu plat)

One cultivar can be pollinated by a compatible cultivar from the same group or close (A with A, or A with B, but not A with C or D).

Cultivars are sometimes classified by the day of peak bloom in the average 30-day blossom period, with pollenizers selected from cultivars within a 6-day overlap period.

Maturation and harvest

Cultivars vary in their yield and the ultimate size of the tree, even when grown on the same rootstock. Some cultivars, if left unpruned, grow very large—letting them bear more fruit, but making harvesting more difficult. Depending on tree density (number of trees planted per unit surface area), mature trees typically bear 40–200 kg (90–440 lb) of apples each year, though productivity can be close to zero in poor years. Apples are harvested using three-point ladders that are designed to fit amongst the branches. Trees grafted on dwarfing rootstocks bear about 10–80 kg (20–180 lb) of fruit per year. Farms with apple orchards open them to the public so consumers can pick their own apples.

Crops ripen at different times of the year according to the cultivar. Cultivar that yield their crop in the summer include 'Gala', 'Golden Supreme', 'McIntosh', 'Transparent', 'Primate', 'Sweet Bough', and 'Duchess'; fall producers include 'Fuji', 'Jonagold', 'Golden Delicious', 'Red Delicious', 'Chenango', 'Gravenstein', 'Wealthy', 'McIntosh', 'Snow', and 'Blenheim'; winter producers include 'Winesap', 'Granny Smith', 'King', 'Wagener', 'Swayzie', 'Greening', and 'Tolman Sweet'.

Storage

Commercially, apples can be stored for a few months in controlled atmosphere chambers to delay ethylene-induced ripening. Apples are commonly stored in chambers with higher concentrations of carbon dioxide and high air filtration. This prevents ethylene concentrations from rising to higher amounts and preventing ripening from occurring too quickly.

For home storage, most cultivars of apple can be held for approximately two weeks when kept at the coolest part of the refrigerator (i.e. below 5 °C). Some can be stored up to a year without significant degradation. Some varieties of apples (e.g. 'Granny Smith' and 'Fuji') have more than three times the storage life of others.

Non-organic apples may be sprayed with a substance 1-methylcyclopropene blocking the apples' ethylene receptors, temporarily preventing them from ripening.

Pests and diseases

Apple trees are susceptible to a number of fungal and bacterial diseases and insect pests. Many commercial orchards pursue a program of chemical sprays to maintain high fruit quality, tree health, and high yields. These prohibit the use of synthetic pesticides, though some older pesticides are allowed. Organic methods include, for instance, introducing its natural predator to reduce the population of a particular pest.

A wide range of pests and diseases can affect the plant. Three of the more common diseases or pests are mildew, aphids, and apple scab.

- Mildew is characterized by light grey powdery patches appearing on the leaves, shoots and flowers, normally in spring. The flowers turn a creamy yellow color and do not develop correctly. This can be treated similarly to Botrytis—eliminating the conditions that caused the disease and burning the infected plants are among recommended actions.
- Aphids are a small insect. Five species of aphids commonly attack apples: apple grain aphid, rosy apple aphid, apple aphid, spirea aphid, and the woolly apple aphid. The aphid species can be identified by color, time of year, and by differences in the cornicles (small paired projections from their rear). Aphids feed on foliage using needle-like mouth parts to suck out plant juices. When present in high numbers, certain species reduce tree growth and vigor.
- **Apple scab:** Apple scab causes leaves to develop olive-brown spots with a velvety texture that later turn brown and become cork-like in texture. The disease also affects the fruit, which also develops similar brown spots with velvety or cork-like textures. Apple scab is spread through fungus growing in old apple leaves on the ground and spreads during warm spring weather to infect the new year's growth.

Among the most serious disease problems is a bacterial disease called fireblight, and two fungal diseases: Gymnosporangium rust and black spot. Other pests that affect apple trees include Codling moths and apple maggots. Young apple trees are also prone to mammal pests like mice and deer, which feed on the soft bark of the trees, especially in winter. The larvae of the apple clearwing moth (red-belted clearwing) burrow through the bark and into the phloem of apple trees, potentially causing significant damage.

Production

World production of apples in 2020 was 86 million tonnes, with China producing 47% of the total (table). Secondary producers were the United States and Turkey.

Nutrition

A raw apple is 86% water and 14% carbohydrates, with negligible content of fat and protein (table). A reference serving of a raw apple with skin weighing 100 grams provides 52 calories and a moderate content of dietary fiber. Otherwise, there is low content of micronutrients, with the Daily Values of all falling below 10%.

Uses

All parts of the fruit, including the skin, except for the seeds, are suitable for human consumption. The core, from stem to bottom, containing the seeds, is usually not eaten and is discarded.

Apples can be consumed in various ways: juice, raw in salads, baked in pies, cooked into sauces and spreads like apple butter, and other baked dishes.

Apples are sometimes used as an ingredient in savory foods, such as sausage and stuffing.

Several techniques are used to preserve apples and apple products. Apples can be canned, dried or frozen. Canned or frozen apples are eventually baked into pies or other cooked dishes. Apple juice or cider is also bottled. Apple juice is often concentrated and frozen.

Popular uses

Apples are often eaten raw. Cultivars bred for raw consumption are termed dessert or table apples.

- In the UK, a toffee apple is a traditional confection made by coating an apple in hot toffee and allowing it to cool. Similar treats in the U.S. are candy apples (coated in a hard shell of crystallized sugar syrup) and caramel apples (coated with cooled caramel).
- Apples are eaten with honey at the Jewish New Year of Rosh Hashanah to symbolize a sweet new year.

Apples are an important ingredient in many desserts, such as apple pie, apple crumble, apple crisp and apple cake. When cooked, some apple cultivars easily form a puree known as apple sauce. Apples are also made into apple butter and apple jelly. They are often baked or stewed and are also (cooked) in some meat dishes. Dried apples can be eaten or reconstituted (soaked in water, alcohol or some other liquid).

Apples are milled or pressed to produce apple juice, which may be drunk unfiltered (called apple cider in North America), or filtered. Filtered juice is often concentrated and frozen, then reconstituted later and consumed. Apple juice can be fermented to make cider (called hard cider in North America), ciderkin, and vinegar. Through distillation, various alcoholic beverages can be produced, such as applejack, Calvados, and apfelwein.

Organic production

Organic apples are commonly produced in the United States. Due to infestations by key insects and diseases, organic production is difficult in Europe. The use of pesticides containing chemicals, such as sulfur, copper, microorganisms, viruses, clay powders, or plant extracts (pyrethrum, neem) has been approved by the EU Organic Standing Committee to improve organic yield and quality. A light coating of kaolin, which forms a physical barrier to some pests, also may help prevent apple sun scalding.

Phytochemicals

Apple skins and seeds contain various phytochemicals, particularly polyphenols which are under preliminary research for their potential health effects.

Non-browning apples

The enzyme polyphenol oxidase causes browning in sliced or bruised apples, by catalyzing the oxidation of phenolic compounds to o-quinones, a browning factor. Browning reduces apple taste, color, and food value. Arctic apples, a non-browning group of apples introduced to the United States market in 2019, have been genetically modified to silence the expression of polyphenol oxidase, thereby delaying a browning effect and improving apple eating quality. The US Food and Drug Administration in 2015, and Canadian Food Inspection Agency in 2017, determined that Arctic apples are as safe and nutritious as conventional apples

Other products

Apple seed oil is obtained by pressing apple seeds for manufacturing cosmetics.

Allergy

One form of apple allergy, often found in northern Europe, is called birch-apple syndrome and is found in people who are also allergic to birch pollen. Allergic reactions are triggered by a protein in apples that is similar to birch pollen, and people affected by this protein can also develop allergies to other fruits, nuts, and vegetables. Reactions, which entail oral allergy syndrome (OAS), generally involve itching and inflammation of the mouth and throat, but in rare cases can also include life-threatening anaphylaxis. This reaction only occurs when raw fruit is consumed—the allergen is neutralized in the cooking process. The variety of apple, maturity and storage conditions can change the amount of allergen present in individual fruits. Long storage times can increase the amount of proteins that cause birch-apple syndrome.

In other areas, such as the Mediterranean, some individuals have adverse reactions to apples because of their similarity to peaches. This form of apple allergy also includes OAS, but often has more severe symptoms, such as vomiting, abdominal pain and urticaria, and can be life-threatening. Individuals with this form of allergy can also develop reactions to other fruits and nuts. Cooking does not break down the protein causing this particular reaction, so affected individuals cannot eat raw or cooked apples. Freshly harvested, over-ripe fruits tend to have the highest levels of the protein that causes this reaction.

Breeding efforts have yet to produce a hypoallergenic fruit suitable for either of the two forms of apple allergy.

Toxicity of seeds

Apple seeds contain small amounts of amygdalin, a sugar and cyanide compound known as a cyanogenic glycoside. Ingesting small amounts of apple seeds causes no ill effects, but consumption of extremely large doses can cause adverse reactions. It may take several hours before the poison takes effect, as cyanogenic glycosides must be hydrolyzed before the cyanide ion is released. The United States National Library of Medicine's Hazardous Substances Data Bank records no cases of amygdalin poisoning from consuming apple seeds.

6

HORTICULTURE OF BERRY

A berry is a small, pulpy, and often edible fruit. Typically, berries are juicy, rounded, brightly colored, sweet, sour or tart, and do not have a stone or pit, although many pips or seeds may be present. Common examples are strawberries, raspberries, blueberries, blackberries, red currants, white currants and blackcurrants. In Britain, soft fruit is a horticultural term for such fruits.

In common usage, the term "berry" differs from the scientific or botanical definition of a fruit produced from the ovary of a single flower in which the outer layer of the ovary wall develops into an edible fleshy portion (pericarp). The botanical definition includes many fruits that are not commonly known or referred to as berries, such as grapes, tomatoes, cucumbers, eggplants, bananas, and chili peppers. Fruits commonly considered berries but excluded by the botanical definition include strawberries, raspberries, and blackberries, which are aggregate fruits and mulberries, which are multiple fruits. Watermelons and pumpkins are giant berries that fall into the category "pepos". A plant bearing berries is said to be bacciferous or baccate.

Berries are eaten worldwide and often used in jams, preserves, cakes, or pies. Some berries are commercially important. The berry industry varies from country to country as do types of berries cultivated or growing in the wild. Some berries such as raspberries and strawberries have been bred for hundreds of years and are distinct from their wild counterparts, while other berries, such as lingonberries and cloudberries, grow almost exclusively in the wild.

While many berries are edible, some are poisonous to humans, such as deadly nightshade and pokeweed. Others, such as the white mulberry, red mulberry, and elderberry, are poisonous when unripe, but are edible when ripe.

Berries have been valuable as a food source for humans since before the start of agriculture, and remain among the primary food sources of other primates. They were a seasonal staple for early hunter-gatherers for thousands of years, and wild berry gathering remains a popular activity in Europe and North America today. In time, humans learned to store berries so that they could be used in the winter. They may be made into fruit preserves, and among Native Americans, mixed with meat and fats as pemmican.

Berries also began to be cultivated in Europe and other countries. Some species of blackberries and raspberries of the genus Rubus have been cultivated since the 17th century, while smooth-skinned blueberries and cranberries of the genus Vaccinium have been cultivated in the United States for over a century. In Japan, between the 10th and 18th centuries, the term ichibigo ichigo referred to many berry crops. The most widely cultivated berry of modern times, however, is the strawberry, which is produced globally at twice the amount of all other berry crops combined.

The strawberry was mentioned by ancient Romans, who thought it had medicinal properties, but it was then not a staple of agriculture. Woodland strawberries began to be grown in French gardens in the 14th century. The musk strawberry (F. moschata), also known as the hautbois strawberry, began to be grown in European gardens in the late 16th century. Later, the Virginia strawberry was grown in Europe and the United States. [when?] The most commonly consumed strawberry, the garden strawberry (F. ananassa), is an accidental hybrid of the Virginia strawberry and a Chilean variety Fragaria chiloensis.

It was first noted by a French gardener around the mid 18th century that, when F. moschata and F. virginiana were planted in between rows of F. chiloensis, the Chilean strawberry would bear abundant and unusually large fruits. Soon after, Antoine Nicolas Duchesne began to study the breeding of strawberries and made several discoveries crucial to the science of plant breeding, such as the sexual reproduction of strawberry. Later, in the early 1800s, English breeders of strawberry made varieties of F. ananassa which were important in strawberry breeding in Europe, and hundreds of cultivars have since been produced through the breeding of strawberries.

Cultivation

Strawberries have been grown in gardens in Europe since the 14th century. Blueberries were domesticated starting in 1911, with the first commercial crop in 1916. Huckleberries of all varieties are not fully domesticated, but domestication was attempted from 1994 to 2010 for the economically significant western huckleberry. Many other varieties of Vaccinium are likewise not domesticated, with some being of commercial importance.

Agricultural methods

Like most other food crops, berries are commercially grown, with both conventional pest management and integrated pest management (IPM) practices. Organically certified berries are becoming more widely available.

Many soft fruit berries require a period of temperatures between 0 °C (32 F) and 10 °C (50 F) for breaking dormancy. In general, strawberries require 200–300 hours, blueberries 650–850 hours, blackberries 700 hours, raspberries 800–1700 hours, currants and gooseberries 800–1500 hours, and cranberries 2000 hours. However, too low a temperature will kill the crops: blueberries do not tolerate temperatures below −29 °C (20.2 F), raspberries, depending on variety, may tolerate as low as −31 °C (-23.8 F), and blackberries are injured below −20 °C (-4 F).

Spring frosts are, however, much more damaging to berry crops than low winter temperatures. Sites with moderate slopes (3–5%), facing north or east, in the northern hemisphere, near large bodies of water, which regulate spring temperature, are considered ideal in preventing spring frost injury to the new leaves and flowers. All berry crops have shallow root systems. Many land-grant university extension offices suggest that strawberries should not be planted more than five years on the same site, due to the danger of black root rot (though many other illnesses go by the same name), which is controlled in major commercial production by annual methyl bromide fumigation.

Besides the number of years in production, soil compaction, the frequency of fumigation, and herbicide usages increase the appearance of black root rot in strawberries. Raspberries, blackberries, strawberries, and many other berries are susceptible to verticillium wilt. Blueberries and cranberries grow poorly if the clay or silt content of the soil is more than 20%, while most other berries tolerate a wide range of soil types. For most berry crops, the ideal soil is well drained sandy loam, with a pH of 6.2–6.8 and a moderate to high organic content; however, blueberries have an ideal pH of 4.2–4.8 and can be grown on muck soils, while blueberries and cranberries prefer poorer soils with lower cation exchange, lower calcium, and lower levels of phosphorus.

Growing most berries organically requires the use of proper crop rotation, the right mix of cover crops, and the cultivation of the correct beneficial microorganisms in the soil. As blueberries and cranberries thrive in soils that are not hospitable to most other plants, and conventional fertilizers are toxic to them, the primary concern when growing them organically is bird management.

Postharvest small fruit berries are generally stored at 90–95% relative humidity and 0 °C. Cranberries, however, are frost sensitive, and should be stored at 3 C (37.4 F).

Blueberries are the only berries that respond to ethylene, but flavor does not improve after harvest, so they require the same treatment as other berries. Removal of ethylene may reduce disease and spoilage in all berries. Precooling within one to two hours post-harvest to storage temperature, generally 0 °C, via forced air cooling increases the storage life of berries by about a third. Under optimum storage conditions, raspberries and blackberries last for two to five days, strawberries 7–10 days, blueberries two to four weeks, and cranberries two to four months. Berries can be shipped under high carbon dioxide or modified atmosphere of 10–15% carbon dioxide for high carbon dioxide or 15–20% carbon dioxide and 5–10% oxygen for a modified atmosphere container to increase shelf life and prevent grey mold rot.

Breeding

Several discoveries in the science of breeding berries were made in the 18th century by Antoine Nicolas Duchesne in his work on strawberries. In the traditional technique of plant breeding, berries with specific desirable characteristics were chosen and allowed to sexually reproduce with other berries, and offspring with improved traits could then be selected and used for further crossing. Plants may be hybridized with different species within the same genus; hybridization between different genus may also be possible, but more difficult. Breeding may seek to increase the size and yield of the fruit, improve the flavor and quality of its nutrient content, such as antioxidants, expand the harvest season, and produce cultivars with resistance to diseases, tolerance of hot or cold conditions, and other desirable traits.

Advancements in molecular biology and genetic engineering allow for a more efficient and better targeted approach in the selection for a desirable genotype, via marker-assisted selection, for example. Genetic modification techniques can also be used for breeding berries.

Horticultural soft fruit berries

Some fruit not commonly referred to as berries and not always botanically berries are included by land-grant university extension offices in their guides for berry cultivation, or in guides for identifying local wild edible and non-edible berries. Examples include beach plums, American persimmons, pawpaws, Pacific crabapples, and prickly pears.

Commercial production

In the year 2005, there were 1.8 million acres of land worldwide cultivating berries, with 6.3 million tons produced.

Economics

In certain regions, berrypicking can be a large part of the economy, and it is becoming increasingly common for western European countries such as Sweden and Finland to import cheap labor from Thailand or Bulgaria to do the berry picking. This practice has come under scrutiny in the past years because of the low wages and poor living standard for the "berry-pickers", as well as the lack of worker safety.

In the late 2010s in the US, reduced migration from Mexico and increased minimum wage standards have made finding "stoop-work" labourers to pick the strawberry crop difficult and costly.

Phytochemicals and color

Once ripened, berries have a contrasting color to their background (often of green leaves), making them visible and attractive to frugivorous animals and birds. This assists the wide dispersal of the plants' seeds.

Berry colors are due to natural phytochemicals, including plant pigments, such as anthocyanins, together with other flavonoids localized mainly in berry skins, seeds and leaves. Although berry pigments have antioxidant properties in vitro, there is no physiological evidence established to date that berry pigments have actual antioxidant or any other functions within the human body. Consequently, it is not permitted to claim that foods containing polyphenols have antioxidant health value on product labels in the United States or Europe.

Culinary significance

Use in baked goods

Berries are commonly used in pies or tarts, such as blueberry pie, blackberry pie, and strawberry pie.

Berries are often used in baking, such as blueberry muffins, blackberry muffins, berry cobblers, berry crisps, berry cakes, berry buckles, berry crumb cakes, berry tea cakes, and berry cookies. Berries are commonly incorporated whole into the batter for baking, and care is often taken so as to not burst the berries. Frozen or dried berries may be preferable for some baked berry products. Fresh berries are also often incorporated into baked berry desserts, sometimes with cream, either as a filling to the dessert or as a topping.

Beverages

Berries are often added to water and/or juiced, as in cranberry juice, which accounts for 95% of cranberry crop usage, blueberry juice, raspberry juice, goji berry juice, acai juice, aronia berry juice, and strawberry juice. Wine is the principal fermented beverage made from berries (grapes). Fruit wines are commonly made out of other berries. In most cases, sugars must be added to the berry juices in the process of Chaptalization to increase the alcohol content of the wine. Examples of fruit wines made from berries include: elderberry wine, strawberry wine, blueberry wine, blackberry wine, redcurrant wine, huckleberry wine, goji wine and cranberry wine. Berries are used in some styles of beer, particularly framboise (made with raspberry) and other fruit lambics.

Dried

Currants, raisins and sultanas are examples of dried grape berries, and many other commercially important berries are available in dried form.

Fruit preserves

Berries are perishable fruits with a short shelf life, and are often preserved by drying, freezing, pickling or making fruit preserves. Berries such as blackberry, blueberry, boysenberry, lingonberry, loganberry, raspberry, and strawberry are often used in jams and jellies. In the United States, Native Americans were "the first to make preserves from blueberries".

Other usages

Chefs have created quick pickled soft fruit, such as blackberries, strawberries, and blueberries. Strawberries can be battered and quickly fried in a deep fryer. Sauces made from berries, such as cranberry sauce, can be frozen until hard, battered, and deep fried. Cranberry sauce is a traditional food item for Thanksgiving, and similar sauces can be made from many other berries such as blueberries, raspberries, blackberries, and huckleberries.

Cultural significance

Dyeing

Berries have been used in some cultures for dyeing. Many berries contain juices that can easily stain, affording use as a natural dye. For example, blackberries are useful for making dyes, especially when ripe berries can easily release juice to produce a colorfast effect. Rubus berries, such as blackberry, raspberry, black raspberry, dewberry, loganberry, and thimbleberry all produce dye colors.

These were once used by Native Americans. In Hawaii, the native raspberry called 'akala' was used to dye tapa cloth with lavender and pink hues, whereas berries from the dianella lily were used for blue coloration, and berries from the black nightshade were used to produce green coloration.

Blueberry Farming Information

Introduction of Blueberries

Blueberries are perennial flowering plants of the genus "Vaccinium" that are usually grown for their dark berries. These berries can be consumed as raw, added to breakfasts, baked goods, and yogurts (curds) or made into jams. Blueberries have incredible nutrition and health benefits. Blueberries are not only popular but also repeatedly ranked in the U.S. diet as having one of the highest antioxidant properties. These berries are native to North America. In India, Blueberry farming is very limited and there is a huge future potential for commercial blueberry farming due to its excellent health benefits. Blueberries have successfully started growing in India with correct cultivation practices. For sure, Blueberry farming will have a bright future in India.

Health Benefits of Blueberries

The following are the health benefits of Blueberries.

- Blueberries are the excellent source of antioxidant foods.
- Blueberries are low in calories and high in nutrients.
- Blueberries may help in protecting against ageing and cancer.
- Blueberries can help in maintaining brain function.
- Blueberries protect cholesterol in the blood from becoming damaged.
- Blueberries may lower blood pressure.
- Blueberries may help reduce muscle damage after a strenuous workout.
- Blueberries can have anti-diabetic effects.
- Blueberries may help fight urinary tract infections.
- Blueberries may help in preventing heart disease.

Local Names for Blueberries in India:

- Avurinelli (Tamil),
- Fall sa (Telugu),
- Nilabadari (Hindi),
- Njaval (Malayalam),
- Jambhool/Karvand (Marathi),
- Falsa (Gujarati),
- Blueberry (Oriya).

Climate Requirement for Blueberry Farming:

Blueberries can be grown on a wide range of climatic conditions. However, they best thrive in warmer (full sun) climates. When you are ready for farming, you should check with nearest horticulture department for suitable cultivar for your area climatic conditions.

Soil Requirement for Blueberry Farming:

Blueberry crop prefers soils which are highly acidic, fertile, aerated, moist and well-drained. The optimal soil pH range is 4.0 to 5.5 for best growth and yield. In case higher pH, adding a small amount of sulfur to the soil can reduce the pH of the soil. Before starting a blueberry cultivation, go for soil testing.

Varieties of Blueberries:

There are many varieties available in Blueberry farming. Every cultivar will produce for a period of 3 to 4 harvesting weekly. However, thy are grouped into 3 categories. : highbush, lowbush and hybrid half-high. The following are the improved varieties of blueberries: Duke, Toro, Chandler, Chanticleer, Oneal, Misty Nelson, Legacy, Elliott, Elizabeth, Arlen, Revielle Prince, Columbus Premier, Powder blue Climax, Bright well, BlueCrop, Blue Ray, BlueJay.

- Propagation in Blueberry Farming:
- Propagation can be done through seeds, hardwood cuttings.
- Land Preparation, Planting in Blueberry Farming:

Land should be levelled and ploughed till it attains the fine tilth stage. Main field should be made weed free. Plant spacing should be 80 cm in between rows and 3 meter aisles. Blueberry planting can be done any time during the year provided there is sufficient irrigation available. 1 year or 2 year plants grown in a 1 liter or 3.5 liter container should be planted in the main field. The preferred stem length is 15 to 25 cm & 25 to 45 cm. Pits should be dug 10 inch deep just before 2 weeks of planting the trees. Pits should be dug out a square about one meter across to give the side-spreading roots. The soil taken out from the pits should be mixed with equal parts of leaf mould, coco peat or compost.

Pruning in Blueberry Farming:

Generally, Blueberry plants are a bush type and grow stems from its crown. Usually, the number of productive stems should be 9 to 12. Pruning should be carried out every year by removing canes that are 5 to 6 years old. One should not allow the blueberry bush to produce fruits for the first couple of years. Pinch back blossoms, this will help to stimulate the growth of the plant. Pruning should be carried out in late winter, preferably just before growth begins. During the initial 4 years period, there is no need of pruning.

Plant protection in Blueberry Farming:

Generally, Blueberry plants are resistant to pests and diseases. The main problem is, these berries are a favourite food for birds, so avoid birds in the orchard by adding nets around the plants. Especially this is needed at the time of fruiting (in June).

Irrigation in Blueberry Farming:

Irrigate the plants immediately after transplanting in the field. Blueberry plants should be irrigated once a week. Rain water is better than the tap water as it is more alkaline in nature. In long dry spell weather conditions, it may require frequent irrigation based on soil moisture holding capacity.

Intercultural Operations in Blueberry Farming: Weeding should be carried regularly by removing dried or damaged leaves. To control the weeds, plant basins should be kept clean and use any kind mulch (in spring season) along with any fertilizer made for acid loving trees. Mulching will save the water loss, controls the weed growth, protect the soil from erosion and mulching material will become well rotten organic fertilizer. Mulching can be done with a 2 to 4 inch layer of wood-chips or saw dust after planting.

Remove basal growth as half of the fruiting and flowers starting at the basin, remove all fruits that are growing on the lower portion of the plants to make the plants stronger especially to make the root system stronger.

Manures and Fertilizers in Blueberry Farming: The use of composts in blueberry farming has many potential benefits. Well-rotten farm yard manure (FMY) like cow dung should be supplemented at the time of land/soil preparation. As Blueberry plants love acidic soils, look for a high acid blueberry bush fertilizers that contain ammonium sulfate, ammonium nitrate, or sulfur-coated urea. These tend to have a lower pH (higher acid). Fertilizers should be applied in the spring before their leaves have grown in.

Harvesting in Blueberry Farming: Blueberry plants will start producing fruits from the second or third season of growth. The plants produce berries once a year. Harvested fruits can be marketed as fresh as well as canned (frozen). After harvesting the berries, all canes hat produced berries should be removed. Usually, harvesting will start in the month of Aug thru Sept. Don't pick the Blueberries as soon as they turn blue and wait for a couple of days. When they are ready to harvest, they should fall off right into your hand.

The yield in Blueberry Farming: The fruit yield depends on many factors like variety (cultivar), soil type, irrigation and weather conditions. Generally, in Blueberry farming, one can expect 1 kg of fruits per each plant in the first harvest. In subsequent years, Blueberry plant will double in yield until the 6th to 7th year of maturity. The maximum yield can be expected form a plant is 10 kg and an average yield is 5 to 6 kg per plant. In most cases, Blueberry plants produce fruits up to 20 to 25 years.

7

FLOWER FARMING

Floriculture or flower farming is the study of growing and marketing flowers and foliage plants. Floriculture includes cultivation of flowering and ornamental plants for direct sale or for use as raw materials in cosmetic and perfume industry and in the pharmaceutical sector. It also includes production of planting materials through seeds, cuttings, budding and grafting. In simpler terms floriculture can be defined as the art and knowledge of growing flowers to perfection. The persons associated with this field are called floriculturists.

Worldwide more than 140 countries are involved in commercial Floriculture. The leading flower producing country in the world is Netherlands and Germany is the biggest importer of flowers. Countries involved in the import of flowers are Netherlands, Germany, France, Italy and Japan while those involved in export are Colombia, Israel, Spain and Kenya. USA and Japan continue to be the highest consumers.

Floriculture in India

Floriculture is an age old farming activity in India having immense potential for generating gainful self-employment among small and marginal farmers. In the recent years it has emerged as a profitable agri-business in India and worldwide as improved standards of living and growing consciousness among the citizens across the globe to live in environment friendly atmosphere has led to an increase in the demand of floriculture products in the developed as well as in the developing countries worldwide. The production and trade of floriculture has increased consistently over the last 10 years. In India, Floriculture industry comprises flower trade, production of nursery plants and potted plants, seed and bulb production, micro propagation and extraction of essential oils. The country has exported 15,695.31 MT of floriculture products to the world for the worth of Rs. 575.98 Crores/77.84 USD Millions in 2020-21.

The Government of India identified floriculture as a sunrise industry and accorded it 100 percent export oriented status. Floriculture products mainly consist of cut flowers, pot plants, cut foliage, seeds bulbs, tubers, rooted cuttings and dried flowers or leaves. The important floricultural crops in the international cut flower trade are rose, carnation, chrysanthemum, gerbera, gladiolus, orchids, anthurium, tulip and lilies. Floriculture crops like gerberas, carnation, etc. are grown in green houses. The open field crops are chrysanthemum, roses, gaillardia, lily marygold, aster, tuberose etc.

According to Horticulture Crops for 2018-19 statistics, the total area under flower crops was 303 thousand hectares. Total area under floriculture in India is second largest in the world and only next to China. Production of flowers was estimated to be 2910 thousand MT of which loose flowers accounted to 2263 thousand MT and cut flowers to 647 thousand MT. Fresh and Dried cut flowers dominate floriculture exports from India.

Among states, in terms of production, Karnataka stands firts with a total of 253.24 thousand tonnes. In terms of area, Kerala is the leader in floriculture with about 53.26 thousand hectares under floriculture cultivation. Other major flower growing states are Tamil Nadu and Andhra Pradesh in the South, West Bengal in the East, Maharashtra in the West and Rajasthan, Delhi and Haryana in the North.

The expert committee set up by Govt. of India for promotion of export oriented floriculture units has identified Bangalore, Pune, New Delhi and Hyderabad as the major areas suitable for such activity especially for cut flowers. Of the four zones identified as potential centers for flower production namely Bangalore, Hyderabad, Pune and New Delhi, the area around Bangalore and Pune have got the advantage of ideal climatic conditions where the temperature ranges between 15 to 30°C. In view of this, the units established in these locations do not require either cooling or heating system. As a result maximum number of units has been established in these locations. There are more than 300 export oriented units in India. APEDA (Agricultural and Processed Food Products Export Development Authority) is the registering authority for such units.

Marketing

In India, marketing of cut flowers is much unorganized. In most of the Indian cities flowers are brought to wholesale markets, which mostly operate in open yards. From here the flowers are distributed to the local retail outlets which more often than not operate in the open on-road sides, with different flowers arranged in large buckets. In the metropolitan cities, however, there are some good florist show rooms, where flowers are kept under controlled temperature conditions, with considerable attention to value added service.The government is now investing in setting up of auction platforms, as well as organized florist shops with better storage facilities to prolong shelf life. The packaging and transportation of flowers from the farms to the retail markets at present is very unscientific. The flowers, depending on the kind, are packed in gunny bags, bamboo baskets, simple cartons or just wrapped in old newspapers and transported to markets by road, rail or by air. However, the government has provided some assistance for buying refrigerated cargos and built up a large number of export oriented units with excellent facilities of pre-cooling chambers, cold stores and reefer vans.

Export Constraints

In spite of an abundant and varied production base, India's export of floricultural product is not encouraging. The low performance is attributed to many constraints like non-availability of air space in major airlines. The Indian floriculture industry is facing with a number of challenges mainly related to trade environment, infrastructure and marketing issues such as high import tariff, low availability of perishable carriers, higher freight rates and inadequate refrigerated and transport facilities. At the production level the industry is faced with challenges mostly related to availability of basic inputs including quality seeds and planting materials, efficient irrigation system and skilled manpower. In order to overcome these problems, steps must be taken to reduce import duty on planting material and equipment, reduce airfreight to a reasonable level, provide sufficient cargo space in major airlines and to establish model nurseries for supplying genuine planting material. Training centres should be established for training the personnel in floriculture and allied areas. Exporters should plan and monitor effective quality control measures right from production to post harvesting,storage, and transportation.

Government Programmes and Policies

Department of Agriculture and Cooperation under the Ministry of Agriculture is the nodal organization responsible for development of the floriculture sector. It is responsible for formulation and implementation of national policies and programmes aimed at achieving rapid agricultural growth through optimum utilization of land, water, soil and plant resources of the country. Production of cut flowers for exports is also a thrust area for support. The Agricultural and Processed Food Products Export Development Authority (APEDA), the nodal organization for promotion of agri exports including flowers, has introduced several schemes for promoting floriculture exports from the country. These relate to development of infrastructure, packaging, market development, subsidy on airfreight for export of cut flowers and tissue-cultured plants, database up-gradation etc. The 100% Export Oriented Units are also given benefits like duty free imports of capital goods. Import duties have also been reduced on cut flowers, flower seeds, tissue-cultured plants, etc. Setting up of walk in type cold storage has been allowed at the International airports for storage of export produce.

Initiatives have also been launched for the benefit of exporters by providing cold storage and cargo handling facility for perishable products at various international airports. Direct subsidy up to 50 percent is also available in cold storage units. Besides, subsidy is also provided by APEDA on improved packaging materials to promote their use. To attract entrepreneurship in floriculture sector, NABARD is providing financial assistance to hi-tech units at reasonable interest rates.

Several schemes have been initiated by the Government for promotion and development of the floriculture sector including "Integrated Development of Commercial Floriculture" which aims at improvement in production and productivity of traditional as well as cut flowers through availability of quality planting material, production of off season and quality flowers through protected cultivation, improvement in post harvest handling of flowers and training persons for a scientific floriculture. Many state governments have set up separate departments for promotion of floriculture in their respective states.

Research work on floriculture is being carried out at several research institutions under the Indian Council of Agricultural Research and Council of Scientific and Industrial Research, in the horticulture/floriculture departments of State Agricultural Universities and under the All India Coordinated Floriculture Improvement Project with a network of about twenty (20) centres. The key focus areas are crop improvement, standardization of agro-techniques including improved propagation methods, plant protection and post harvest management. In recent years, however, technologies for protected cultivation and tissue culture for mass propagation have also received attention. A large number of promising varieties of cut flowers have been developed. All these efforts indicate the government's commitment for improving the sector and creating a positive environment for entrepreneurship development in the field.

Demand and Supply

The demand for flowers is seasonal as it is in most countries. The demand for flowers has two components: a steady component and a seasonal component. The factors which influence the demand are to some extent different for traditional and modern flowers.

Traditional Flowers: The steady demand for traditional flowers comes from the use of flowers for religious purposes, decoration of homes and for making garlands and wreaths. This demand is particularly strong in Kerala, Karnataka, Tamil Nadu, Odisha and West Bengal, as the use of flowers for above mentioned purposes is part of their local culture. The bulk of seasonal demand comes from festivals and marriages. The demand is generally for specific flowers.

Modern Flowers: The bulk of the steady demand for modern flowers comes from institutions like hotels, guest houses and marriage gardens. The demand is concentrated in urban areas. With increasing modernization and globalization the demand for modern flowers from the individual consumers is likely to grow enormously as the trend of "say it with flowers" is increasing and the occasions which call for flower giving will continue to present themselves. Although there is an increasing demand for modern flowers from individuals, institutions continue to be the dominant buyers in the market. The price of these flowers also depends on their demand and varies accordingly.

Green House Technology for Flower Production

In present scenario of increasing demand for cut flowers protected cultivation in green houses is the best alternative for using land and other resources more efficiently. In protected environment suitable environmental conditions for optimum plant growth are provided which ultimately provide quality products. Green House is made up of glass or plastic film, which allows the solar radiations to pass through but traps the thermal radiations emitted by plants inside and thereby provide favourable climatic conditions for plant growth. It is also used for controlling temperature, humidity and light intensity inside. On the basis of basic material used, building cost and technology used, green houses can be of three types-

Low-cost greenhouse: The low-cost green house is made of polythene sheet of 700 gauge supported on bamboos with twines and nails. Its size depends on the purpose of its utilization and availability of space. The temperature within greenhouse increases by 6-100C more than outside.

Medium-cost greenhouse: With a slightly higher cost greenhouse can be framed with GI pipe of 15 mm bore. This greenhouse has a covering of UV -stabilized polythene of 800 gauge. The exhaust fans are used for ventilation which are thermostatically controlled. Cooling pad is used for humidifying the air entering the chamber. The greenhouse frame and glazing material have a life span of about 20 years and 2 years respectively.

Hi-tech greenhouse: In this type of green house the temperature, humidity and light are automatically controlled according to specific plant needs. These are indicated through sensor or signal-receiver. Sensor measures the variables, compare the measurement to a standard value and finally recommend to run the corresponding device. Temperature control system consists of temperature sensor heating/cooling mechanism and thermostat operated fan. Similarly, relative humidity is sensed through optical tagging devices. Boiler operation, irrigation and misting systems are operated under pressure sensing system. This modern structure is highly expensive, requiring qualified operators, maintenance, care and precautions. However, these provide best conditions for export quality cut flowers and are presently used by large number of export units.

Floriculture has emerged as an important agribusiness, providing employment opportunities and entrepreneurship in both urban and rural areas. National Horticulture Board helps one to establish a flower business. Agricultural and Processed Food Products Export Development Authority helps entrepreneurs with cold storage facilities and freight subsidies. It has been found that Commercial Floriculture has higher potential per unit area than most of the field crops and therefore a lucrative business. During the last decade there has been a thrust on export of cut flowers. The export surplus has found its way into the local market influencing people in cities to purchase and use flowers in their daily lives. Floriculture thus, offers a great opportunity to farmers in terms of income generation and empowerment. Small and marginal farmers may also use every inch of their land for raising the flower and foliage crops. Floriculture also offers careers in production, marketing, export and research. One can find employment in the floriculture industry as a farm manager, plantation expert, supervisor or project coordinator. Besides, one can work as consultant or landscape architect with proper training. In addition, floriculture also provides career opportunities in service sector which include such jobs as floral designers, landscape designers, landscape architects and horticultural therapists. Research and teaching are some other avenues of employment in the field.

Chrysanthemum

Chrysanthemums, sometimes called mums or chrysanths, are flowering plants of the genus Chrysanthemum in the family Asteraceae. They are native to Asia and northeastern Europe. Most species originate from East Asia and the center of diversity is in China. There are countless horticultural varieties and cultivars.

Taxonomy

The genus once included more species, but was split several decades ago into several genera, putting the economically important florist's chrysanthemums in the genus Dendranthema. The naming of the genera has been contentious, but a ruling of the International Botanical Congress in 1999 changed the defining species of the genus to Chrysanthemum indicum, restoring the florist's chrysanthemums to the genus Chrysanthemum.

The other species previously included in the narrow view of the genus Chrysanthemum are now transferred to the genus Glebionis. The other genera separate from Chrysanthemum include Argyranthemum, Leucanthemopsis, Leucanthemum, Rhodanthemum, and Tanacetum.

Description

Wild Chrysanthemum taxa are herbaceous perennial plants or subshrubs. They have alternately arranged leaves divided into leaflets with toothed or occasionally smooth edges. The compound inflorescence is an array of several flower heads, or sometimes a solitary head. The head has a base covered in layers of phyllaries. The simple row of ray florets are white, yellow or red; many horticultural specimens have been bred to bear many rows of ray florets in a great variety of colors. The disc florets of wild taxa are yellow. The fruit is a ribbed achene. Chrysanthemums, also known as 'mums', are one of the prettiest varieties of perennials that start blooming early in the fall. This is also known as favorite flower for the month of November.

Chrysanthemums were first cultivated in China as a flowering herb as far back as the 15th century BC.Over 500 cultivars had been recorded by the year 1630. The plant is renowned as one of the Four Gentlemen in Chinese and East Asian art. The plant is particularly significant during the Double Ninth Festival. The flower may have been brought to Japan in the eighth century AD, and the Emperor adopted the flower as his official seal. The "Festival of Happiness" in Japan celebrates the flower.

Chrysanthemums entered American horticulture in 1798 when Colonel John Stevens imported a cultivated variety known as 'Dark Purple' from England. The introduction was part of an effort to grow attractions within Elysian Fields in Hoboken, New Jersey.

Economic uses

Ornamental Uses

Modern cultivated chrysanthemums are showier than their wild relatives. The flower heads occur in various forms, and can be daisy-like or decorative, like pompons or buttons. This genus contains many hybrids and thousands of cultivars developed for horticultural purposes. In addition to the traditional yellow, other colors are available, such as white, purple, and red. The most important hybrid is Chrysanthemum × morifolium (syn. C. × grandiflorum), derived primarily from C. indicum, but also involving other species.

Over 140 varieties of chrysanthemum have gained the Royal Horticultural Society's Award of Garden Merit.

Chrysanthemums are divided into two basic groups, garden hardy and exhibition. Garden hardy mums are new perennials capable of wintering in most northern latitudes. Exhibition varieties are not usually as sturdy. Garden hardies are defined by their ability to produce an abundance of small blooms with little if any mechanical assistance, such as staking, and withstanding wind and rain. Exhibition varieties, though, require staking, overwintering in a relatively dry, cool environment, and sometimes the addition of night lights.

The exhibition varieties can be used to create many amazing plant forms, such as large disbudded blooms, spray forms, and many artistically trained forms, such as thousand-bloom, standard (trees), fans, hanging baskets, topiary, bonsai, and cascades.

Chrysanthemum blooms are divided into 10 different bloom forms by the US National Chrysanthemum Society, Inc., which is in keeping with the international classification system. The bloom forms are defined by the way in which the ray and disk florets are arranged. Chrysanthemum blooms are composed of many individual flowers (florets), each one capable of producing a seed. The disk florets are in the center of the bloom head, and the ray florets are on the perimeter. The ray florets are considered imperfect flowers, as they only possess the female productive organs, while the disk florets are considered perfect flowers, as they possess both male and female reproductive organs.

Irregular incurves are bred to produce a giant head called an ogiku. The disk florets are concealed in layers of curving ray florets that hang down to create a 'skirt'. Regular incurves are similar, but usually with smaller blooms and a dense, globular form. Intermediate incurve blooms may have broader florets and a less densely flowered head.

In the reflex form, the disk florets are concealed and the ray florets reflex outwards to create a mop-like appearance. The decorative form is similar to reflex blooms, but the ray florets usually do not radiate at more than a 90° angle to the stem.

The pompon form is fully double, of small size, and very globular in form. Single and semidouble blooms have exposed disk florets and one to seven rows of ray florets.

In the anemone form, the disk florets are prominent, often raised and overshadowing the ray florets. The spoon-form disk florets are visible and the long, tubular ray florets are spatulate.

In the spider form, the disk florets are concealed, and the ray florets are tube-like with hooked or barbed ends, hanging loosely around the stem. In the brush and thistle variety, the disk florets may be visible.

Culinary uses

Yellow or white chrysanthemum flowers of the species C. morifolium are boiled to make a sweet drink in some parts of Asia. The resulting beverage is known simply as chrysanthemum tea. In Korea, a rice wine flavored with chrysanthemum flowers is called gukhwaju.

Chrysanthemum leaves are steamed or boiled and used as greens, especially in Chinese cuisine. The flowers may be added to dishes such as mixian in broth, or thick snakemeat soup to enhance the aroma. Small chrysanthemums are used in Japan as a sashimi garnish.

Insecticidal uses

Pyrethrum (Chrysanthemum [or Tanacetum] cinerariaefolium) is economically important as a natural source of insecticide. The flowers are pulverized, and the active components, called pyrethrins, which occur in the achenes, are extracted and sold in the form of an oleoresin. This is applied as a suspension in water or oil, or as a powder. Pyrethrins attack the nervous systems of all insects, and inhibit female mosquitoes from biting. In sublethal doses they have an insect repellent effect. They are harmful to fish, but are far less toxic to mammals and birds than many synthetic insecticides. They are not persistent, being biodegradable, and also decompose easily on exposure to light. Pyrethroids such as permethrin are synthetic insecticides based on natural pyrethrum.

Environmental uses

Chrysanthemum plants have been shown to reduce indoor air pollution by the NASA Clean Air Study.

Rose

A rose is a woody perennial flowering plant of the genus Rosa, in the family Rosaceae, or the flower it bears. There are over a hundred species and thousands of cultivars. They form a group of plants that can be erect shrubs, climbing or trailing with stems that are often armed with sharp prickles. Flowers vary in size and shape and are usually large and showy, in colours ranging from white through yellows and reds. Most species are native to Asia, with smaller numbers native to Europe, North America, and northwestern Africa. Species, cultivars and hybrids are all widely grown for their beauty and often are fragrant. Roses have acquired cultural significance in many societies. Rose plants range in size from compact, miniature roses, to climbers that can reach seven meters in height. Different species hybridize easily, and this has been used in the development of the wide range of garden roses.

The name rose comes from French, itself from Latin rosa, which was perhaps borrowed from Oscan, from Greek ñüäïí rhódon (Aeolic âñüäïí wródon), itself borrowed from Old Persian wrd- (wurdi), related to Avestan varYäa, Sogdian ward, Parthian wâr.

Botany

The leaves are borne alternately on the stem. In most species they are 5 to 15 centimetres (2.0 to 5.9 in) long, pinnate, with (3–) 5–9 (–13) leaflets and basal stipules; the leaflets usually have a serrated margin, and often a few small prickles on the underside of the stem. Most roses are deciduous but a few (particularly from South east Asia) are evergreen or nearly so.

The flowers of most species have five petals, with the exception of Rosa sericea, which usually has only four. Each petal is divided into two distinct lobes and is usually white or pink, though in a few species yellow or red. Beneath the petals are five sepals (or in the case of some Rosa sericea, four). These may be long enough to be visible when viewed from above and appear as green points alternating with the rounded petals. There are multiple superior ovaries that develop into achenes.

Roses are insect-pollinated in nature. The aggregate fruit of the rose is a berry-like structure called a rose hip. Many of the domestic cultivars do not produce hips, as the flowers are so tightly petalled that they do not provide access for pollination. The hips of most species are red, but a few (e.g. Rosa pimpinellifolia) have dark purple to black hips. Each hip comprises an outer fleshy layer, the hypanthium, which contains 5–160 "seeds" (technically dry single-seeded fruits called achenes) embedded in a matrix of fine, but stiff, hairs. Rose hips of some species, especially the dog rose (Rosa canina) and rugosa rose (Rosa rugosa), are very rich in vitamin C, among the richest sources of any plant. The hips are eaten by fruit-eating birds such as thrushes and waxwings, which then disperse the seeds in their droppings. Some birds, particularly finches, also eat the seeds.

While the sharp objects along a rose stem are commonly called "thorns", they are technically prickles — outgrowths of the epidermis (the outer layer of tissue of the stem). (True thorns, as produced by e.g. Citrus or Pyracantha, are modified stems, which always originate at a node and which have nodes and internodes along the length of the thorn itself.) Rose prickles are typically sickle-shaped hooks, which aid the rose in hanging onto other vegetation when growing over it. Some species such as Rosa rugosa and Rosa pimpinellifolia have densely packed straight prickles, probably an adaptation to reduce browsing by animals, but also possibly an adaptation to trap wind-blown sand and so reduce erosion and protect their roots (both of these species grow naturally on coastal sand dunes). Despite the presence of prickles, roses are frequently browsed by deer. A few species of roses have only vestigial prickles that have no points.

Species

Further information: List of Rosa species

The genus Rosa is subdivided into four subgenera:

- Hulthemia (formerly Simplicifoliae, meaning "with single leaves") containing one or two species from southwest Asia, R. persica and Rosa berberifolia which are the only roses without compound leaves or stipules.
- Hesperrhodos (from the Greek for "western rose") contains Rosa minutifolia and Rosa stellata, from North America.
- Platyrhodon (from the Greek for "flaky rose", referring to flaky bark) with one species from east Asia, Rosa roxburghii (also known as the chestnut rose).
- Rosa (the type subgenus, sometimes incorrectly called Eurosa) containing all the other roses. This subgenus is subdivided into 11 sections.
- white and yellow flowered roses from China.
- **Bracteatae** – three species, two from China and one from India.
- **Caninae** – pink and white flowered species from Asia, Europe and North Africa.
- **Carolinae** – white, pink, and bright pink flowered species all from North America.
- **Chinensis** – white, pink, yellow, red and mixed-color roses from China and Burma.
- **Gallicanae** – pink to crimson and striped flowered roses from western Asia and Europe.
- **Gymnocarpae** – one species in western North America (Rosa gymnocarpa), others in east Asia.
- **Laevigatae** – a single white flowered species from China
- **Pimpinellifoliae** – white, pink, bright yellow, mauve and striped roses from Asia and Europe.
- **Rosa (syn. sect. Cinnamomeae)** – white, pink, lilac, mulberry and red roses from everywhere but North Africa.
- **Synstylae** – white, pink, and crimson flowered roses from all areas.

Uses

Roses are best known as ornamental plants grown for their flowers in the garden and sometimes indoors. They have been also used for commercial perfumery and commercial cut flower crops. Some are used as landscape plants, for hedging and for other utilitarian purposes such as game cover and slope stabilization. They also have minor medicinal uses.

Ornamental plants

The majority of ornamental roses are hybrids that were bred for their flowers. A few, mostly species roses are grown for attractive or scented foliage (such as Rosa glauca and Rosa rubiginosa), ornamental thorns (such as Rosa sericea) or for their showy fruit (such as Rosa moyesii).

Ornamental roses have been cultivated for millennia, with the earliest known cultivation known to date from at least 500 BC in Mediterranean countries, Persia, and China. Many thousands of rose hybrids and cultivars have been bred and selected for garden use as flowering plants. Most are double-flowered with many or all of the stamens having mutated into additional petals.

In the early 19th century the Empress Josephine of France patronized the development of rose breeding at her gardens at Malmaison. As long ago as 1840 a collection numbering over one thousand different cultivars, varieties and species was possible when a rosarium was planted by Loddiges nursery for Abney Park Cemetery, an early Victorian garden cemetery and arboretum in England.

Cut flowers

Roses are a popular crop for both domestic and commercial cut flowers. Generally they are harvested and cut when in bud, and held in refrigerated conditions until ready for display at their point of sale.

In temperate climates, cut roses are often grown in glasshouses, and in warmer countries they may also be grown under cover in order to ensure that the flowers are not damaged by weather and that pest and disease control can be carried out effectively. Significant quantities are grown in some tropical countries, and these are shipped by air to markets across the world.

Some kind of roses are artificially coloured using dyed water, like rainbow roses.

Perfume

Rose perfumes are made from rose oil (also called attar of roses), which is a mixture of volatile essential oils obtained by steam distilling the crushed petals of roses. An associated product is rose water which is used for cooking, cosmetics, medicine and religious practices. The production technique originated in Persia and then spread through Arabia and India, and more recently into eastern Europe. In Bulgaria, Iran and Germany, damask roses (Rosa × damascena 'Trigintipetala') are used. In other parts of the world Rosa × centifolia is commonly used. The oil is transparent pale yellow or yellow-grey in colour. 'Rose Absolute' is solvent-extracted with hexane and produces a darker oil, dark yellow to orange in colour. The weight of oil extracted is about one three-thousandth to one six-thousandth of the weight of the flowers; for example, about two thousand flowers are required to produce one gram of oil.

Gaillardia

Gaillardia /aejÈlQĐrdiY/ (common name blanket flower) is a genus of flowering plants in the sunflower family, Asteraceae, native to North and South America. It was named after an M. Gaillard de Charentonneau, an 18th-century French magistrate who was a patron of botany. The common name may refer to the resemblance of the inflorescence to the brightly patterned blankets made by Native Americans, or to the ability of wild taxa to blanket the ground with colonies. Many cultivars have been bred for ornamental use.

Description

These are annual or perennial herbs or subshrubs, sometimes with rhizomes. The stem is usually branching and erect to a maximum height around 80 centimeters. The leaves are alternately arranged. Some taxa have only basal leaves. They vary in shape. They are glandular in most species. The inflorescence is a solitary flower head. The head can have 15 or more ray florets, while some taxa lack any ray florets. They can be almost any shade of yellow, orange, red, purplish, brown, white, or bicolored. They are sometimes rolled into a funnel shape. There are many tubular disc florets at the center of the head in a similar range of colors, and usually tipped with hairs. The fruit usually has a pappus of scales.

Ecology

Gaillardia species are used as food plants by the caterpillars of some Lepidoptera species, including Schinia bina (which has been recorded on G. pulchella), Schinia masoni (which feeds exclusively on G. aristata) and Schinia volupia (which feeds exclusively on G. pulchella).

Gerbera

Gerbera L. is a genus of plants Asteraceae (daisy family). It was named in honour of German botanist and medical doctor Traugott Gerber | (1710-1743) who travelled extensively in Russia and was a friend of Carl Linnaeus.

Gerbera is native to tropical regions of South America, Africa and Asia. The first scientific description of a Gerbera was made by J.D. Hooker in Curtis's Botanical Magazine in 1889 when he described Gerbera jamesonii, a South African species also known as

Transvaal daisy or Barberton Daisy. Gerbera is also commonly known as the African Daisy.

Gerbera species bear a large capitulum with striking, two-lipped ray florets in yellow, orange, white, pink or red colours. The capitulum, which has the appearance of a single flower, is actually composed of hundreds of individual flowers. The morphology of the flowers varies depending on their position in the capitulum. The flower heads can be as small as 7 cm (Gerbera mini 'Harley') in diameter or up to 12 cm (Gerbera 'Golden Serena').

Gerbera is very popular and widely used as a decorative garden plant or as cut flowers. The domesticated cultivars are mostly a result of a cross between Gerbera jamesonii and another South African species Gerbera viridifolia. The cross is known as Gerbera hybrida. Thousands of cultivars exist. They vary greatly in shape and size. Colours include white, yellow, orange, red, and pink. The centre of the flower is sometimes black. Often the same flower can have petals of several different colours. Gerbera is also important commercially. It is the fifth most used cut flower in the world (after rose, carnation, chrysanthemum, and tulip). It is also used as a model organism in studying flower formation.

Gerbera contains naturally occurring coumarin derivatives. Gerbera is a tender perennial plant. It is attractive to bees, butterflies and/or birds, but resistant to deer. Their soil should be kept moist but not soaked.

Marigold

Calendula is a genus of about 15–20 species of annual and perennial herbaceous plants in the daisy familyAsteraceae that are often known as marigolds. They are native to southwestern Asia, western Europe, Macaronesia, and the Mediterranean. Other plants are also known as marigolds, such as corn marigold, desert marigold, marsh marigold, and plants of the genus Tagetes. The genus name Calendula is a modern Latin diminutive of calendae, meaning "little calendar", "little clock" or possibly "little weather-glass". The common name "marigold" refers to the Virgin Mary. The most commonly cultivated and used member of the genus is the pot marigold (Calendula officinalis). Popular herbal and cosmetic products named 'calendula' invariably derive from C. officinalis.

Traditional and ancient uses

Calendula species have been used traditionally as culinary and medicinal herbs. The petals are edible and can be used fresh in salads or dried and used to color cheese or as a replacement for saffron. A yellow dye has been extracted from the flowers.

Romans and Greeks used the golden calendula in many rituals and ceremonies, sometimes wearing crowns or garlands made from the flowers. One of its nicknames is "Mary's Gold," referring to the flowers' use in early Catholic events in some countries. Calendula flowers are sacred flowers in India and have been used to decorate the statues of Hindu deities since early times.

Calendula ointments are skin products used to treat minor cuts, burns, and skin irritation.

Chemistry

The flowers of C. officinalis contain flavonol glycosides, triterpene oligoglycosides, oleanane-type triterpene glycosides, saponins, and a sesquiterpene glucoside.

Pharmacological effects

Calendula oil is still used medicinally. The oil of C. officinalis is used as an anti-inflammatory, an antitumor agent, and a remedy for healing wounds.

Plant pharmacological studies have suggested that Calendula extracts have antiviral, antigenotoxic, and anti-inflammatory properties in vitro. In herbalism, Calendula in suspension or in tincture is used topically for treating acne, reducing inflammation, controlling bleeding, and soothing irritated tissue. Limited evidence indicates Calendula cream or ointment is effective in treating radiation dermatitis. Topical application of C. officinalis ointment has helped to prevent dermatitis and pain; thus reducing the incidence rate of skipped radiation treatments in randomized trials.

Calendula has been used traditionally for abdominal cramps and constipation. In experiments with rabbit jejunum, the aqueous-ethanol extract of C. officinalis flowers was shown to have both spasmolytic and spasmogenic effects, thus providing a scientific rationale for this traditional use. An aqueous extract of C. officinalis obtained by a novel extraction method has demonstrated antitumor (cytotoxic) activity and immunomodulatory properties (lymphocyte activation) in vitro, as well as antitumor activity in mice.

Calendula plants are known to cause allergic reactions, and should be avoided during pregnancy.

Culinary use

Calendula species have been used in cooking for centuries. The flowers were a common ingredient in German soups and stews, which explains the nickname "pot marigold". The lovely golden petals were also used to add color to butter and cheese. The flowers are traditional ingredients in Mediterranean and Middle Eastern dishes. Calendula tea provides health benefits, as well as being delicious.

Dye use

The beautiful flowers were once used as a source of dye for fabrics. By using different mordants, a variety of yellows, oranges and browns could be obtained.

Medicinal use

Ancient cultures recognized and used the healing properties of calendula. In some of the earliest medical writings, calendula was recommended for treating ailments of the digestive tract. It was used to detoxify the liver and gall bladder. The flowers were applied to cuts and wounds to stop bleeding, prevent infection and speed healing. Calendula was also used for various women's ailments, and to treat a number of skin conditions. During the American Civil War, calendula flowers were used on the battlefields in open wounds as antihemorrhagic and antiseptic, and they were used in dressing wounds to promote healing. Calendula also was used in this way during World War I. Calendula has been historically significant in medicine in many cultures, and it is still important in alternative medicine today.

Flower Farming Business Plan or Floriculture Business Plan in India

Flower Cultivated Areas in India

Karnataka, Tamil Nadu, and Andhra Pradesh, West Bengal, Maharashtra, and Rajasthan, Haryana is the major flower cultivated states in India. Flower cultivation has emerged as a profitable agri-business in recent years and worldwide as improved standards of living. Other flower-producing states like Madhya Pradesh, Jharkhand, Mizoram, Orissa, Gujarat, Haryana, Assam, and Chhattisgarh. In simpler terms, Flower farming can be defined as the art and knowledge of growing flowers to perfection. The persons associated with this flower farming field are known as floriculturists.

Scope of Flower Cultivation in India

The scope of floriculture in India is as follows;

- The opportunities for flower cultivation are increasing due to several uses of flowers in an aesthetic sense, urbanization, and substantially increased purchasing power of people.
- The demand for flowering plants and their products, such as bouquets, garlands, and value-added products like dry flowers and potpourris is increasing by the day. They are required in several functions and celebrations.
- The diverse agro-climatic conditions in the country enable the growth of all types of flowers in one season or the other seasons.
- Landscaping has become an integral component of urban horticulture, which apart from adding aesthetic value to a place, protects the environment, reduces air and also noise pollution, and promotes ecotourism.
- In light of climate change, the scope of turf or lawn grasses, vertical gardening, and roof gardening, etc., is rising.

- Increasing industrialization and depleting agricultural land have opened avenues for potted plant production and marketing.

Requirements for Successful Flower Farming Business

The successful commercial flower farming process mostly depends on the varieties of flowers. In the latest technology, there are many planting methods are available but plant cultivation in the greenhouse method is very famous and convenient for flower plants. Flower plants can be growing in pots, on the terrace, indoors, open fields, greenhouse, and polyhouse.

Floriculture is referred to as growing and marketing flower plants and these pants are largely produced in plant-growing structures in temperate climates. It is largely thought of as a greenhouse industry, though several flowers are cultivated outdoors in nurseries or crop fields.

Cut flowers are a perfect cash crop because the plants are very easy to grow, produce quickly, and can supply a good income throughout the growing season. Also, startup costs are low, as most gardeners already have the basic gardening tools required, and only have to buy seeds to get started. The cut flower farming business is known to be a profitable business in India; it has evolved from the backyard garden to a global industry in all countries where it's carried out.

Several people attracted to the floriculture business because of their interest in growing flowers. Though, actual plant production accounts for only a portion of what is required for a successful business;

- Grow a quality product
- Sell the crop
- Handle / Store / Transport the crop
- Production efficiency
- Marketing
- Business management
- Make a profit

Flower cultivation should be equally adept at marketing their crops and managing the business to make a profit. Some considerations must be made before starting a flower farming business.

Flower Seed and Planting Material

There is a high demand for quality flower seeds particularly annual ornamentals and ornamental planting material. Flower seeds of annuals are formed in huge quantities for sale. A large number of bulbous flower plants like Gladiolus, Tuberose, Amaryllis, Dahlia, Lilies, Freesia, Tulip, and Calla Lily, etc., are marketed.

When sowing flower seeds, you need to have a good potting mix and also need a warm area to germinate the seeds. Sow the flower seeds by sprinkling them over the damp potting mix and then very lightly with more mix. It takes most perennial seeds 3 to 5 weeks to germinate so you do need to be patient. Frequently some of the flower seeds from the same variety will germinate faster than others so you will have a few seeds up and growing while others are still dormant. Then, place the seedlings into good light and let them grow. Do not expect all the seeds to germinate.

For perennial plants, the germination rate can be as low as 50% in some plant varieties, compared to almost 95% of annual seed that will germinate. Also, you can increase the germination rate by chilling the seeds before you sow them and soaking them. The young perennial will grow steadily the 1st year but they do not always flower that first year. Growing perennials from seed are different from growing annuals, but it is just as worthwhile.

Flower Farming Business Ideas in India

Ageratum

Ageratum plant is a fast-growing annual plant. Caring for the ageratum plant includes regular watering until the plant is established. With successive plantings, it can produce continuous blooms from early summer to frost. Make sure soil drains well and don't let the soil dry out either. Ageratum plants quickly wilt if conditions are too dry.

Alstroemeria

Growing alstroemeria plant is easy and does not need much time or effort. You can find at least 50 species of alstroemeria around the globe. Alstroemeria flowers have no fragrance and it has a vase life of about two weeks. The alstroemeria plant needs full sun and well-drained soil for farming.

Carnations

Carnation is popular as "The Flowers of God". The flower comes in numerous colors. You can grow carnation from cuttings. Also, the flowers need some hours of full sun each day. Keep the soil in moist condition. However, avoid overwatering the plant. The plants growing in an area that gets 4 to 6 hours of sunlight per day.

Gerbera

Gerberas come in different colors. Also, it is one of the most popular cut flowers globally and the flower has a long vase life. The most inexpensive method to produce gerberas is from seed. However, you should obtain the seed from reputed seed suppliers. Gerbera plants are perennials and do best in full sun, in well-drained soil. If soil is poorly drained, grow the plants in a raised garden bed.

Jasmine

The scientific name of jasmine is Jasminum and the flower belongs to the Oleaceae family. You can find several fresh flower usages like making garlands, bouquets, decorating hair of women, and religious offerings, etc. Well-drained, rich loamy soil with a pH level from 6.5-7.5 is ideal for their cultivation. It prefers a mild and tropical climate. Jasmine is commercially cultivated in India under open field conditions.

Lavender

The Lavender plant needs full sun to do well, but not too much summer heat. Also, select the right lavender variety that suits your locality. The growing lavender plant is easy and rewarding. To grow a lavender plant you could need well-drained soil.

Orchid

Generally, there are two growth types of orchids. The first is known as Monopodial Orchids and the second known as Sympodial Orchids. Generally, the orchid plant needs good sunlight. You will require providing sun lights 12 to 14 hours each day.

Rose

Rose is a perennial shrub or vine of the Rosa genus and belongs to the Rosaceae family. Generally, the rose plant grows well in the plains under the ideal condition of fertile loamy soils with salt-free irrigation water. A moderately cool climate with bright sunshine and free ventilation is necessary for rose cultivation in India. The ideal soil for rose plants should be medium loam having sufficient organic matter, proper drainage, and a pH level of 6.0 to 7.0. You can harvest the Rose flowers at the tight bud stage when the color is fully developed.

Sunflowers

Sunflowers are the early-blooming varieties at less than 60 days to harvest and are popular market sellers. Sunflower plants grow well in the soil with lots of water holding capacity, drainage, and fertilizer.

Sunflower Farming (pic source: pixabay)

Tulip

The tulip plant prefers full sun but will settle for less. The Tulip flowers good revenue return. Also, the flowers have good export potential.

Peony

The peony is a flowering plant in the Paeonia genus, the only genus in the family Paeoniaceae. There are hundreds of varieties of peony perennial plants. Good air circulation around the plant is important. Peonies love to be in full sun and will flower at their best in that position, but they will tolerate light shade.

Zinnias

Nothing fills out a bouquet quite like zinnia flowers, with their brilliant colors, shapes, and sizes. The giant varieties are bestsellers at the market. Zinnia plants grow and flower best in full sun.

Violets

Violets have heart-shaped leaves and asymmetrical flowers. For violet cultivation, you will need to select a site with full sun to light shade. Provide water only moderately. Although they love cool conditions, it doesn't need huge amounts of water.

Tips for Starting Flower Farming Business in India

Here some tips to follow to start a flower farming business;

Floriculture is a great opportunity for people who love gardening and also earn good money from the flower farming business. To understand the flower farming business you want to understand the term floriculture and what it consists of for example pot plants, cut foliage, seeds bulbs, tubers, rooted cuttings and dried flowers, cut flowers, or leaves.

Choose Planting Site Carefully

Plant the right plant in the right place is important for the flower farming business. Provide full sun for flower plants that need it and shade for those that need shade. Start in a small space and then get bigger as you get more comfortable. If you start a garden too big, you might get discouraged.An area with too much direct sun or too much shade will be difficult for some flower plants to grow in. Find out what type of light is best for the flower plants.

You may end up wanting to select an area with more or less sun than your original plot has. If you plan on planting several different types of flowers, choose ones with similar light or shade requirements so that they grow equally well in the same location.

Prepare the Soil

Flower cultivation starts with healthy soil. Generally, most flowering plants do best in soil that's loose and well-drained with a lot of organic material in it. You don't want to dig a large area to plant flowers, but you must dig enough soil that you can add some compost to improve the soil structure and add nutrients. Also, add organic matter to allow air, water, worms, and roots to roam freely.

If your flower plants are struggling, poor soil is usually the culprit. Regardless of whether you are planting your flower plants in a pot or a garden, good soil is a must. Find out what potting medium is the best for the flower plants you are growing and then try to find the best location or potting mix for your flowers.

Fertilize Plants Sparingly

A few perennial plants are "heavy feeders" and they require more nutrients than other perennial plants to bloom and grow well. Lilies, tall phlox, and delphiniums are examples of perennial flower plants that need a little more fertilizer than most.

However, most perennial plants don't need much fertilizer and can perform poorly if over-fertilized. Generally, a single spring application of a balanced fertilizer like 5-5-5 or 10-10-10 works for the heavy feeders; just scratch a handful into the soil around the base of the plant. An inch or two of aged manure or finished compost applied in the fall or spring season before mulch is replaced will keep other plants happy and will continue to improve the soil.

Water your Flowers Regularly

Although individual plant needs vary based on humidity and the type of plant, it is common to add several cups of water to each plant by using a watering close to the soil. Also, you can have a sprinkler or drip system installed to do the work for you on an automated basis.

Plan your business

Before starting any business you need planning. To keep your flower fresh and increase its life you need a freezer and what other aspects you need to keep in mind. Also, you should analyze the manpower requirement that you might need to design the floral arrangement and for delivery. Either you want to go for a franchise or source to get your business like florists or flower companies. Get proper mentoring before starting the flower farming business as this business requires a lot of knowledge.

Your Flower Farm Should Be Organic

Your products must be natural, which means you use natural pesticides instead of chemicals to get rid of pests. These chemicals could damage your flowers instead of helping them. It is easy to use a natural pesticide solution instead of opting to buy the chemical ones. Use natural predators like praying mantises and ladybugs. Then, these insects prey on the natural pests of plants, so with them on your farm, you have natural and effective pest control.

Choose the Flowers You Want To Cultivate

Your main aim is to earn a profit in the flower farming business. Therefore, select the flowers that are in great demand in the market. You want to segregate them accordingly to help your customers make an informed decision as well as to provide them value for money.

High Demand Flowers in India

The most commercially used flowers in India are Rose, Gerbera, Tulip, Carnation, Chrysanthemum, Orchid, Lilium, Gladiolus, Jasmine, Anthurium, and Hydrangea, etc.

Cultivation Methods to Profit from Flower Cultivation

Open Farming is good for flower farming but there are a few requirements to open field farming. High maintenance, labor, and the possibility of pests and other problems related to weather conditions. A controlled environment can be a great way to increase profits in flower cultivation. The degree of control will determine the investment for your farming project. There are several options to creating a controlled environment. To begin with, you could have a simple greenhouse, but even creating a greenhouse with a drip irrigation system is more helpful. This will need that you choose the right plant, Get the right customer, market the product, and sell the produce at a high price when ready.

Greenhouse for Flower Cultivation in India

Greenhouse for Rose Plantation and most of the flower cultivation, especially for Cut flowers is essential. While Rose plants cultivated in the open field have the capacity of 3.75 lakh stems to 5 Lakh stems per year, cultivation in the greenhouse system can yield 4.75 Lakh to 8 Lakh per acre every year. You could say it's nearly double in quantity and whets even better is the fact that the roses cultivated in greenhouses were of better quality and perfect for export. With better quality also came better prices and resulted in more profit. Common Flowers grown in Greenhouses are Gerbera, Orchid, Anthurium, Lily, Rose, and Carnation.

Market Demand and Supply of Flowers in India

The demand for flowers is seasonal and it has mainly two components. They are a steady component and a seasonal component. The factors which influence the demand are different for traditional and modern flower plants.

Traditional Flowers – The steady demand for traditional flowers comes from the use of flowers for religious purposes, and decoration purposes. This demand is mainly strong in Kerala, Karnataka, Tamil Nadu, Odisha, and West Bengal. The bulk of seasonal demand flowers come from festivals and marriages. The demand is normally for specific flowers.

Modern Flowers – The bulk of the demand for modern flowers comes from institutions like hotels, guest houses, and marriage gardens. The demand is concentrated in urban areas. The price of these flowers also depends on their demand and changes accordingly.

Steps to Follow for Flower Farming Business Plan

Step 1) Making a business plan is important in starting a flower farming business. Decide the mission statement and business objective. According to the availability of flowers in your location or area, you will need to select what types of flower products you will be creating.

Step 2) Then, specify your target niche, marketing, and advertising strategy. Also, you will want to calculate the startup budget including store set up, employee, and raw materials. Calculate an expected return on investment. A proper flower farming business plan will help you to take the right steps at the right time and to get finance.

Step 3) The main step is to determine if you have the right personality to own and operate your own business. Visit with people who already own their flower businesses. Analyze your strengths and weaknesses. You need to be able to do everything well to succeed in your business.

Step 4) Obtain sufficient capital to start the flower business and operate until it makes a profit. Unless you are quite lucky, your flower farming business will not make money the first year and may not for several years. Generally, flower production operations do not require much initial capital but some major expenses including irrigation, greenhouses for production, and hoop houses.

Step 5) Determine if there is a sufficient market for the products or services your flower business will offer. Define the location you intend to reach. Direct retail will depend on people who can readily drive or walk to your business, which means that you must be in or close to a metropolitan area.

Step 6) Your potential customer base may be large enough but may not be able to support your flower business if there are too many competitors already in the market. Analyze your competition and remember lower prices will not work in the long run. That strategy only works for large companies who can produce or sell large amounts of products and also services and make money through economy of scale.

Step 7) The primary customers are those who you anticipate will buy most of the products or services. They are the ones that you will direct most of the marketing towards. Secondary customers are those who can purchase some of your products and services but not in large enough amounts to support the business. You should always consider secondary customers, especially in case your primary customers fail to support your flower business, but do not pay so much attention to secondary customers that you lose your primary customers. For example, a cut flower grower can target florists as their primary customers and have wholesalers or a grocery store as their secondary customers.

Most Common Problems in Growing Flower Plants

Common Pests and their control

Spider mites, mealy bugs, fungus gnats, whitefly, and aphids are the most common pests in flowering plants. Flower plants that are being moved from decks, patios, etc. to the home or office must be washed thoroughly with a spray of clear water.

Control – A period of isolation (10-14 days) can be useful for controlling these pests. Most of these pests in flowering plants can be controlled by washing off with a spray of water. Chemical sprays are not pleasant to use for indoor flower plants and are recommended for outdoor application.

Diseases and their control

Diseases are rare on flowering plants. Botrytis is the main disease in flowering plants and soft leaf spot or crown rot fungi which thrive in moist atmospheres. The best-preventing method for this disease is to remove affected parts. Also, some other preventing methods are given more space for air circulation or move infected plants to a drier place

8

HORTICULTURE OF FRUIT CROPS

Horticulture sector is an integral element for food, nutritional and economic security in the country. Horticulture involves crop improvement, production, post-harvest management and marketing of fruits, vegetables, aromatic and herbal plants, flowers, spices and plantation crops.

As far as fruit crops are concerned, India is the second largest producer of fruits in the world. The wide range of agro-climatic conditions of India is conducive for growing a large variety of fruits in tropical, subtropical, temperate and arid zones of the country.

Fruit growing is one of the important and paying branches of horticulture. It has been practiced in India since times. The art and science has now developed into one of the most deviled skilful and intensive forms of land utilization. The standard of living of the people of countries is judge by the production and per capita conservation of fruits. Fruit growing has several economic advantages.

Scope and importance of fruit crops

Fruit growing is one of the important and age old practices, practiced in India since ancient times. Cultivation of fruit crops plays an important role in overall status of the mankind and the nation. The standard of living of the people of a country is depending upon the production and per capita consumption of fruits. Fruit growing have more economic advantages.

I. Economic importance

High productivity: High yield per unit area: From a unit area of land more yield is realized from fruit crops than any of the agronomic crops. The average yields of Papaya, Banana and Grapes are 10 to 15 times than that of agronomic crops.

High net profit: Through, the initial cost of establishment of an orchard is high, it is compensated by higher net profit due to higher productivity or high value of produce.

Eg- Wheat/GN/Ragi- 3.0 -4.0 tonnes/ha-25-35,000-00,

Grapes/Mango/Banana-20-40t/ha-1.5-2.5 lakh/ha.

Source of raw material for agro based industries: Fruit farming provides raw materials for various agro based industries- canning and preservation (fresh fruits), coir industries (coconut husk), pharmaceutical industry (Aonla, Papaya, Jamun) Transporting and packaging industries etc.

Efficient utilization of resources: Growing of fruits being perennial in nature, enables grower to remain engaged throughout the year in farm operations and to utilize fully the resources & assets like machinery, labour, land water for production purpose throughout the year compared to agronomic crops.

Utilization of waste and barren lands for production: Although, most of the fruits crops require perennial irrigation and good soil for production, there are many fruit crops of hardy in nature, Mango, Ber, Cashew, Custard apple, Aonla, Phalsa, Jamun etc. which are grown on poor shallow, undulated soils considered unsuitable for growing grain/ agronomical crops.

Foreign exchange: Many fresh fruits, processed products and spices are exported to several countries earning good amount of foreign exchange.

1. Per Unit Yields are Height: Well maintained and established orchards bring better returns than many of the fields crops. From a unit area of land more yield/ income is realized than any of the agronomic crops. The average yields o papaya and banana are 10 to 15 times more than of agronomical crops.

Out of the total area under cultivation grain crops account for 72% with a production o 210 million tons. While due to very high productivity fruit corps yields over 60 million tons from only 0.7 % of the total cultivable area.

2. High Net Profits: Though the initial cost off establishment of an orchard is height. It is compensated by higher productivity or due to high value of produce. In cashew though the average per tree yields is less (4 to 5 kg / plant) due to its market value (200 to 250 Rs / kg) it fetches higher economic returns (800 to 1000 Rs/plant.).

Fruit farming provides raw material for various agrobased industries like canning and preservation (fresh fruits).

1. Efficient Utilization of Resources:

Agronomic crops are seasonal in nature and hence, farmers have to engage themselves in other occupations during slack seasons, but fruit growing being perennial in nature, en abeles grower to remain engaged throughout the year in farm operations and utilize full the resources and assets like machinery, in farm, land, water for production purpose through the year.

2. Utilization of Waste and Barren Lands for Production:

Although most of the fruit crop requires perennial and good soils for production, there are many fruit drops of hardy in nature like mango, ber, cashew, custard apple, anola, phalsa, jamun etc. which are grown on poor, shallow, undulated soils considered unsuitable for growing grain/ agronomical crops. Traditional farming proves uneconomic on waste lands. Development of cashew and mango plantations on hill slopes in Kankan where even ragi or nachani cannot be grown, have brought additional revenge to the growers.

3. Ability of Earning Foreign Exchange:

Many fresh fruits and processed products and spices are exported to several countries earning good amount of foreign exchange. Out of the total exports of agricultural produce of Rs. 2900 million. Horticultural crops which mostly include fruits and its produce contribute Rs.15460 million (52% of total export of argil. Produce). India earns a foreign exchanges of Rs. 277 .4 cores by exporting spices.

1. One Time Capital Investment: Most of the fruit crops are perennial in nature and hence, there is no recurring expenditure o planting and layout of a fruit orchard.

2. Continuous Flow of Money: Harvesting of most of the fruit is of continuous in nature and being and being highly perishable, need to be marketed immediately after harvest. This provides a source of continuous flow of inputs and for other expenses of immediate nature as against agronomic crops which are harvested at home time.

i. Fruit growing in kitchen garden gardens helps to reduce family budget on purchase of fruits.

ii. Planting of trees helps to maintain ecological balance and to increase precipitations of the locality. Fruit trees for such planting programmers also generate economy.

iii. Fruit tree farming also reduce soil erosion, silting tanks and air pollution.

iv. Generate employment: Fruit tree faming being highly intensive and skilful enterprise,

v. Generate employment even for trained persons.

Phyllanthus emblica

Phyllanthus emblica, also known as emblic, emblic myrobalan, myrobalan, Indian gooseberry, Malacca tree, or amlafrom Sanskrit amalika is a deciduous tree of the family Phyllanthaceae. It is known for its edible fruit of the same name.

Plant Morphology and Harvesting

The tree is small to medium in size, reaching 1–8 m (3 ft 3 in–26 ft 3 in) in height. The branchlets aren't glabrous or finely pubescent, 10–20 cm (3.9–7.9 in) long, usually deciduous; the leaves are simple, subsessile and closely set along branchlets, light green, resembling pinnate leaves. The flowers are greenish-yellow. The fruit is nearly spherical, light greenish yellow, quite smooth and hard on appearance, with six vertical stripes or furrows.

Ripening in autumn, the berries are harvested by hand after climbing to upper branches bearing the fruits. The taste of Indian emblic is sour, bitter and astringent, and it is quite fibrous. In India, it is common to eat emblic steeped in salt water and red chilli powder to make the sour fruits palatable.

Cultural and Religious Significance

The tree is considered sacred by Hindus as God Vishnu is believed to dwell in it. The tree is worshiped on Amalaka Ekadashi.

In other Hindu beliefs, Amla is said to have originated from the drops of Amrit which spilled on earth accidentally, because of the fight of Gods and Demons after ksheera sagar manthan. And hence also this religious belief makes claims that it almost cures every disease and is also good in extending the longevity of life.

In the Sanskrit Buddhist tradition half an amalaka fruit was the final gift to the Buddhist sangha by the great Indian emperor Ashoka. This is illustrated in the Ashokavadana in the following verses:

"A great donor, the lord of men, the eminent Maurya Ashoka, has gone from being lord of Jambudvipa [the continent] to being lord of half a myrobalan." (Strong, 1983, p. 99) This deed became so famous that a stupa was created to mark the place of the event in modern-day Patna and was known as the Amalaka stupa.

According to Hindu tradition, Adi Shankara of Kerala composed and recited the Kanakadhara stotram in praise of Mahalakshmi to make a poor Brahmin lady get wealth, in return for a single amla presented to him as bhiksha on an auspicious dwadashi day. Contemporary poet/philosopher Ravi Teja Yelamanchili wrote a book titled Amalaki. The book is based on Advaita Vedanta of Sri Adi Shankaracharya.

According to a Tamil legend, Avvaiyar, a female poet, ethicist and political activist of the Sangam period was gifted with one amla by King Athiyaman to give her long life.

In Theravada Buddhism, this plant is said to have been used as the tree for achieving enlightenment, or Bodhi by twenty first Lord Buddha called "Pussa -

Traditional Uses

Traditional Medicine

In traditional Indian medicine, dried and fresh fruits of the plant are used. All parts of the plant are used in various Ayurvedic/Unani medicine (Jawarish amla) herbal preparations, including the fruit, seed, leaves, root, bark and flowers. According to

Ayurveda, amla fruit is sour (amla) and astringent (kashaya) in taste (rasa), with sweet (madhura), bitter (tikta) and pungent (katu) secondary tastes (anurasas). Its qualities (gunas) are light (laghu) and dry (ruksha), the postdigestive effect (vipaka) is sweet (madhura) and its energy (virya) is cooling (shita).

According to Ayurveda, amla balances all three doshas. While amla is unusual in that it contains five out of the six tastes recognized by Ayurveda, it is most important to recognize the effects of the "virya", or potency, and "vipaka", or post-digestive effect. Considered in this light, amla is particularly helpful in reducing pitta because of its cooling energy. It also balances both Pitta and vata by virtue of its sweet taste. The kapha is balanced primarily due to its drying action.

It may be used as a rasayana (rejuvenative) to promote longevity, and traditionally to enhance digestion (dipanapachana), treat constipation (anuloma), reduce fever (jvaraghna), purify the blood (raktaprasadana), reduce cough (kasahara), alleviate asthma (svasahara), strengthen the heart (hrdaya), benefit the eyes (chakshushya), stimulate hair growth (romasanjana), enliven the body (jivaniya), and enhance intellect (medhya).

In Ayurvedic polyherbal formulations, Indian gooseberry is a common constituent, and most notably is the primary ingredient in an ancient herbal rasayana called Chyawanprash. This formula, which contains 43 herbal ingredients as well as clarified butter, sesame oil, sugar cane juice, and honey, was first mentioned in the Charaka Samhita as a premier rejuvenative compound.

In Chinese traditional therapy, this fruit is called, which is used to treat throat inflammation.

Culinary use

Maharashtra is one of the largest producers and suppliers of Indian gooseberries. In this region the fruit is commonly pickled with salt, oil, and spices. The amla fruit is eaten raw or cooked into various dishes. In Andhra Pradesh, tender varieties are used to prepare dal (a lentil preparation), and amle ka murabbah, a sweet dish made by soaking the berries in sugar syrup until they are candied. It is traditionally consumed after meals.

In the Batak area of Sumatra, Indonesia, the inner bark is used to impart an astringent, bitter taste to the broth of a traditional fish soup known as holat.

Other uses

Popularly used in inks, shampoos and hair oils, the high tannin content of Indian gooseberry fruit serves as a mordant for fixing dyes in fabrics. Amla shampoos and hair oil are traditionally believed to nourish the hair and scalp and prevent premature grey hair. to get clean and sweet water for drinking and cooking.

Pharmacological Research

In vitro (on cells)

Indian gooseberry has undergone preliminary research, demonstrating in vitro antiviral and antimicrobial properties. There is preliminary evidence in vitro that its extracts induce apoptosis and modify gene expression in osteoclasts involved in rheumatoid arthritis and osteoporosis. It may prove to have potential activity against some cancers.

In vivo (in animals)

- One recent animal study found treatment with E. ofûcinalis reduced severity of acute pancreatitis (induced by L-arginine in rats). It also promoted the spontaneous repair and regeneration process of the rat pancreas occurring after an acute attack.
- Experimental preparations of leaves, bark or fruit have shown potential efficacy against laboratory models of disease, such as for inflammation, cancer, age-related renal disease, and diabetes.
- Emblica officinalis tea may ameliorate diabetic neuropathy due to aldose reductase inhibition.

In rats it significantly reduced blood glucose, food intake, water intake and urine output in diabetic rats compared with the non diabetic control group.

Another recent study with alloxan-induced diabetic rats given an aqueous amla fruit extract has shown significant decrease of the blood glucose, as well as triglyceridemic levels and an improvement of the liver function caused by a normalization of the liver-specific enzyme alanine transaminase activity.

Clinical Studies

A human pilot study demonstrated a reduction of blood cholesterol levels in both normal and hypercholesterolemic men with treatment.

Chemical Constituents

Although these fruits are reputed to contain high amounts of ascorbic acid (vitamin C), up to 445 mg per 100 g, the specific contents are disputed, and the overall antioxidant strength of amla may derive instead from its high density of ellagitannins such as emblicanin A (37%), emblicanin B (33%), punigluconin (12%) and pedunculagin (14%). It also contains punicafolin and phyllanemblinin A, phyllanemblin other polyphenols: flavonoids, kaempferol, ellagic acid and gallic acid.

Guava

Guavas (singular guava, /ÈawQĐ.vY/) are common tropical fruits cultivated and enjoyed in many tropical and subtropical regions.

Psidium guajava (common guava, lemon guava) is a small tree in the Myrtle family (Myrtaceae), native to Mexico, Central America, and northern South America. Although related species may also be called guavas, they actually belong to other species or genera, such as the "pineapple guava" Acca sellowiana.

Types

The most frequently eaten species, and the one often simply referred to as "the guava", is the apple guava (Psidium guajava). Guavas are typical Myrtoideae, with tough dark leaves that are opposite, simple, elliptic to ovate and 5–15 centimetres (2.0–5.9 in) long. The flowers are white, with five petals and numerous stamens. The fruits are many-seeded berries. The genera Accara and Acca (formerly Feijoa, pineapple guava) were formerly included in Psidium.

Etymology and Regional Names

The term "guava" appears to derive from Arawak guayabo "guava tree", via the Spanish guayaba. It has been adapted in many European and Asian languages, having a similar form.

Another term for guavas is peru, derived from pear. It is common in countries bordering the western Indian Ocean and probably derives from Spanish or Portuguese. In parts of the Indian subcontinent and Middle-East, guava is called amrood, possibly a variant of armoot meaning "pear" in the Arabic and Turkish languages.

Origin and Distribution

Guavas originated from an area thought to extend from Mexico or Central America and was distributed throughout tropical America and Caribbean region. They were adopted as a crop in subtropical and tropical Asia, the southern United States (from Tennessee and North Carolina south, as well as the west and Hawaii), and tropical Africa.

- Guavas are now cultivated in many tropical and subtropical countries. Several species are grown commercially; apple guava and its cultivars are those most commonly traded internationally.
- Mature trees of most species are fairly cold-hardy and can survive temperatures slightly colder than 25 °F ("4 °C) for short periods of time, but younger plants will likely freeze to the ground.
- Guavas were introduced to Florida in the 19th century and are now grown in Florida as far north as Sarasota, Chipley, Waldo and Fort Pierce. However, they are a primary host of the Caribbean fruit fly and must be protected against infestation in areas of Florida where this pest is present.
- Guavas also grow in a small part of the south of western Europe, specifically the Costa del Sol on Málaga, (Spain) where the guavas are commercially grown since the middle of the 20th century and they proliferate as cultivars.

- Guavas are of interest to home growers in subtropical areas as one of the few tropical fruits that can grow to fruiting size in pots indoors. When grown from seed, guavas bear fruit as soon as two years and as long as 40 years.

Ecology

Psidium species are used as food plants by the caterpillars of some Lepidoptera, mainly moths like the Ello Sphinx (Erinnyis ello), Eupseudosoma aberrans, E. involutum, and Hypercompe icasia. Mites, like Pronematus pruni and Tydeus munsteri, are known to be crop pests of the apple guava (P. guajava) and perhaps other species. The bacterium Erwinia psidii causes rot diseases of the apple guava.

Although the fruit is cultivated and favored by humans, many animals and birds consume it, readily dispersing the seeds in their droppings and, in Hawaii, strawberry guava (P. littorale) has become an aggressive invasive species threatening extinction to more than 100 other plant species. By contrast, several guava species have become rare due to habitat destruction and at least one (Jamaican guava, P. dumetorum), is already extinct.

Guava wood is used for meat smoking in Hawaii and is used at barbecue competitions across the United States. In Cuba and Mexico, the leaves are used in barbecues.

Fruit

Guava fruits, usually 4 to 12 centimetres (1.6 to 4.7 in) long, are round or oval depending on the species. They have a pronounced and typical fragrance, similar to lemon rind but less sharp. The outer skin may be rough, often with a bitter taste, or soft and sweet. Varying between species, the skin can be any thickness, is usually green before maturity, but becomes yellow, maroon, or green when ripe. The pulp inside may be sweet or sour and off-white ("white" guavas) to deep pink ("red" guavas). The seeds in the central pulp vary in number and hardness, depending on species.

Culinary uses

In Mexico, the guava agua fresca beverage is popular. The entire fruit is a key ingredient in punch, and the juice is often used in culinary sauces (hot or cold), as well as artisan candies, dried snacks, fruit bars, desserts, or dipped in chamoy. Pulque de guava is a popular blend of the native alcoholic beverage.

In many countries, guava is eaten raw, typically cut into quarters or eaten like an apple, whereas in other countries it is eaten with a pinch of salt and pepper, cayenne powder or a mix of spices (masala). It is known as the winter national fruit of Pakistan. In the Philippines, ripe guava is used in cooking sinigang. Guava is a popular snack in Taiwan, sold on many street corners and night markets during hot weather, accompanied by packets of dried plum powder mixed with sugar and salt for dipping. In east Asia, guava is commonly eaten with sweet and sour dried plum powder mixtures. Guava juice is popular in many countries. The fruit is also often prepared in fruit salads.

Because of its high level of pectin, guavas are extensively used to make candies, preserves, jellies, jams, and marmalades (such as Brazilian goiabada and Colombian and Venezuelan bocadillo), and also for juices and aguas frescas or may be used in a marmalade jam on toast.

Red guavas can be used as the base of salted products such as sauces, substituting for tomatoes, especially to minimize acidity. A drink may be made from an infusion of guava fruits and leaves, which in Brazil is called chá-de-goiabeira, i.e., "tea" of guava tree leaves, considered medicinal.

Constituents

Nutrients

Guavas are rich in dietary fiber and vitamin C, with moderate levels of folic acid. Having a generally broad, low-calorie profile of essential nutrients, a single common guava (P. guajava) fruit contains about four times the amount of vitamin C as an orange.

However, nutrient content varies across guava cultivars. Although the strawberry guava (P. littorale var. cattleianum) has only 25% of the amount found in more common varieties, its total vitamin C content in one serving (90 mg) still provides 100% of the Dietary Reference Intake.

Phytochemicals

Guava leaves contain both carotenoids and polyphenols like (+)-gallocatechin and leucocyanidin. As some of these phytochemicals produce the fruit skin and flesh color, guavas that are red-orange tend to have more polyphenol and carotenoid content than yellow-green ones.

Mango

The mango is a juicy stone fruit (drupe) belonging to the genus Mangifera, consisting of numerous tropical fruiting trees, cultivated mostly for edible fruit. The majority of these species are found in nature as wild mangoes. They all belong to the flowering plant family Anacardiaceae. The mango is native to South Asia, from where it has been distributed worldwide to become one of the most cultivated fruits in the tropics.

- While other Mangifera species (e.g. horse mango, Mangifera foetida) are also grown on a more localized basis, Mangifera indica—the "common mango" or "Indian mango"—is the only mango tree commonly cultivated in many tropical and subtropical regions.
- It is the national fruit of India, Pakistan, and the Philippines, and the national tree of Bangladesh.

Description

Mango trees grow up to 35–40 m (115–131 ft) tall, with a crown radius of 10 m (33 ft). The trees are long-lived, as some specimens still fruit after 300 years. In deep soil, the taproot descends to a depth of 6 m (20 ft), with profuse, wide-spreading feeder roots; the tree also sends down many anchor roots, which penetrate several feet of soil. The leaves are evergreen, alternate, simple, 15–35 cm (5.9–13.8 in) long, and 6–16 cm (2.4–6.3 in) broad; when the leaves are young they are orange-pink, rapidly changing to a dark, glossy red, then dark green as they mature. The flowers are produced in terminal panicles 10–40 cm (3.9–15.7 in) long; each flower is small and white with five petals 5–10 mm (0.20–0.39 in) long, with a mild, sweet odor suggestive of lily of the valley. Over 400 varieties of mangoes are known, many of which ripen in summer, while some give double crop. The fruit takes three to six months to ripen.

The ripe fruit varies in size and color. Cultivars are variously yellow, orange, red, or green, and carry a single flat, oblong pit that can be fibrous or hairy on the surface, and which does not separate easily from the pulp. Ripe, unpeeled mangoes give off a distinctive resinous, sweet smell. Inside the pit 1–2 mm (0.039–0.079 in) thick is a thin lining covering a single seed, 4–7 cm (1.6–2.8 in) long. The seed contains the plant embryo. Mangoes have recalcitrant seeds; they do not survive freezing and drying.

Etymology

The English word "mango" (plural "mangoes" or "mangos") originated from the Malayalam word mâEEa via Portuguese (also manga) during spice trade with Kerala in 1498. The word's first recorded attestation in a European language was a text by Ludovico di Varthema in Italian in 1510, as manga; the first recorded occurrences in languages such as French and postclassical Latin appear to be translations from this Italian text. The origin of the "-o" ending in English is unclear. Mango is also mentioned by Hendrik van Rheede, the Dutch commander of Malabar (Northern Kerala) in his book Hortus Malabaricus, a compendium of the plants of economic and medical value in the Malabar, published in 1678. When mangoes were first imported to the American colonies in the 17th century, they had to be pickled because of lack of refrigeration. Other fruits were also pickled and came to be called "mangoes", especially bell peppers, and by the 18th century, the word "mango" became a verb meaning "to pickle". In Mandarin Chinese, mango is "',œg", or "mángguÒ" in Pinyin. This is most likely a loanword from English.

Cultivation

Mangoes have been cultivated in South Asia for thousands of years and reached Southeast Asia between the fifth and fourth centuries BCE. By the 10th century CE, cultivation had begun in East Africa. The 14th-century Moroccan traveler Ibn Battuta reported it at Mogadishu. Cultivation came later to Brazil, Bermuda, the West Indies, and Mexico, where an appropriate climate allows its growth.

The mango is now cultivated in most frost-free tropical and warmer subtropical climates; almost half of the world's mangoes are cultivated in India alone, with the second-largest source being China. Mangoes are also grown in Andalusia, Spain (mainly in Málaga province), as its coastal subtropical climate is one of the few places in mainland Europe that permits the growth of tropical plants and fruit trees. The Canary Islands are another notable Spanish producer of the fruit. Other cultivators include North America (in South Florida and California's Coachella Valley), South and Central America, the Caribbean, Hawai'i, south, west, and central Africa, Australia, China, South Korea, Pakistan, Bangladesh, and Southeast Asia. Though India is the largest producer of mangoes, it accounts for less than 1% of the international mango trade; India consumes most of its own production.

Many commercial cultivars are grafted on to the cold-hardy rootstock of Gomera-1 mango cultivar, originally from Cuba. Its root system is well adapted to a coastal Mediterranean climate. Many of the 1,000+ mango cultivars are easily cultivated using grafted saplings, ranging from the "turpentine mango" (named for its strong taste of turpentine) to the huevos de toro. Dwarf or semidwarf varieties serve as ornamental plants and can be grown in containers. A wide variety of diseases can afflict mangoes.

Cultivars

There are many hundreds of named mango cultivars. In mango orchards, several cultivars are often grown in order to improve pollination. Many desired cultivars are monoembryonic and must be propagated by grafting or they do not breed true. A common monoembryonic cultivar is 'Alphonso', an important export product, considered as "the king of mangoes".

Cultivars that excel in one climate may fail elsewhere. For example, Indian cultivars such as 'Julie', a prolific cultivar in Jamaica, require annual fungicide treatments to escape the lethal fungal disease anthracnose in Florida. Asian mangoes are resistant to anthracnose.

The current world market is dominated by the cultivar 'Tommy Atkins', a seedling of 'Haden' that first fruited in 1940 in southern Florida and was initially rejected commercially by Florida researchers. Growers and importers worldwide have embraced the cultivar for its excellent productivity and disease resistance, shelf life, transportability, size, and appealing color. Although the Tommy Atkins cultivar is commercially successful, other cultivars may be preferred by consumers for eating pleasure, such as Alphonso.

Generally, ripe mangoes have an orange-yellow or reddish peel and are juicy for eating, while exported fruit are often picked while underripe with green peels. Although producing ethylene while ripening, unripened exported mangoes do not have the same juiciness or flavor as fresh fruit.

Like other drupaceous fruits, mangoes occur in both freestone and clingstone varieties.

Custard Apple

Casimiroa edulis, with English common names casimiroa, white sapote, and Mexican apple, and known as cochitzapotl in the Nahuatl language (meaning “‘sleep-sapote”) is a species of tropical fruiting tree in the family Rutaceae, native to eastern Mexico and Central America south to Costa Rica. The genus is named for “an Otomi Indian, Casimiro Gómez, from the town of Cardonal in Hidalgo, Mexico, who fought and died in Mexico’s war of independence.”

Description

Mature Casimiroa edulis trees range from 5–16 metres (16–52 ft) tall and are evergreen. The leaves are alternate, palmately compound with 3–5 leaflets, the leaflets 6–13 cm long and 2.5–5 cm broad with an entire margin, and the leaf petiole 10–15 cm long.

The fruit is an ovoid drupe, 5–10 cm in diameter, with a thin, inedible skin turning from green to yellow when ripe, and an edible pulp, which can range in flavor from bland to banana-like to peach to pear to vanilla flan. The pulp can be creamy-white in green skin varieties or a beige-yellow in yellow skin varieties and has a smooth texture similar to ripe avocado. It contains from one to five seeds that are said to have narcotic properties.

Chemical constituents

In the past 40 years, experiments have been carried out on the white sapote’s seeds that have yielded the identity of many pharmacologically active compounds, including: N-methylhistamine, N,N-dimethylhistamine and histamine. It also contains 22 ,5,6-Trimethoxyflavone, 22 ,5,6,7-tetramethoxyflavone (zapotin) and 5-hydroxy-22 ,6,7-trimethoxyflavone (zapotinin).

Health effects

Several recent in vitro studies have shown that zapotin has potential anti-carcinogenic effects against isolated colon cancer cells.

Eating the fruit has long been known to produce drowsiness, as noted by Francisco Hernández de Toledo in the 16th century.Europeans were not the first to note this quality; the Nahuatl name of the plant, cochitzapotl (meaning “‘sleep-sapote”), is a record of the fact that people knew the plant induced drowsiness.

Taxonomy

Unlike the mamey sapote, white sapote is a member of the family Rutaceae, to which citrus belongs. The black sapote is also unrelated and is actually a species of persimmon. This confusion may be because “sapote” comes from the Nahuatl (Aztec) word tzapotl, used to describe all soft, sweet fruit. Commonly grown in Northern New South Wales, Australia, and often mistaken for a persimmon, although these two fruits are unrelated.

Banana

The banana is an edible fruit, botanically a berry, produced by several kinds of large herbaceous flowering plants in the genusMusa. In some countries, bananas used for cooking may be called plantains, in contrast to dessert bananas. The fruit is variable in size, color and firmness, but is usually elongated and curved, with soft flesh rich in starch covered with a rind which may be green, yellow, red, purple, or brown when ripe. The fruits grow in clusters hanging from the top of the plant. Almost all modern edible parthenocarpic bananas come from two wild species – Musa acuminata and Musa balbisiana. The scientific names of most cultivated bananas are Musa acuminata, Musa balbisiana, and Musa × paradisiaca for the hybrid Musa acuminata × M. balbisiana, depending on their genomic constitution. The old scientific name Musa sapientum is no longer used.

Musa species are native to tropical Indomalaya and Australia, and are likely to have been first domesticated in Papua New Guinea.They are grown in 135 countries, primarily for their fruit, and to a lesser extent to make fiber, banana wine and banana beer and as ornamental plants.

Worldwide, there is no sharp distinction between "bananas" and "plantains". Especially in the Americas and Europe, "banana" usually refers to soft, sweet, dessert bananas, particularly those of the Cavendish group, which are the main exports from banana-growing countries. By contrast, Musa cultivars with firmer, starchier fruit are called "plantains". In other regions, such as Southeast Asia, many more kinds of banana are grown and eaten, so the simple two-fold distinction is not useful and is not made in local languages.

The term "banana" is also used as the common name for the plants which produce the fruit. This can extend to other members of the genus Musa like the scarlet banana (Musa coccinea), pink banana (Musa velutina) and the Fe'i bananas. It can also refer to members of the genus Ensete, like the snow banana (Ensete glaucum) and the economically important false banana (Ensete ventricosum). Both genera are classified under the banana family, Musaceae.

Taxonomy

The genus Musa was created by Carl Linnaeus in 1753. The name may be derived from Antonius Musa, physician to the Emperor Augustus, or Linnaeus may have adapted the Arabic word for banana, mauz. The old biological name Musa sapientum = "Muse of the wise" arose because of homophony in Latin with the classical Muses.

Musa is in the family Musaceae. The APG III system assigns Musaceae to the order Zingiberales, part of the commelinid clade of the monocotyledonous flowering plants. Some 70 species of Musa were recognized by the World Checklist of Selected Plant Families as of January 2013; several produce edible fruit, while others are cultivated as ornamentals.

The classification of cultivated bananas has long been a problematic issue for taxonomists. Linnaeus originally placed bananas into two species based only on their uses as food: Musa sapientum for dessert bananas and Musa paradisiaca for plantains. Subsequently, further species names were added. However, this approach proved inadequate to address the sheer number of cultivars existing in the primary center of diversity of the genus, Southeast Asia. Many of these cultivars were given names which proved to be synonyms.

In a series of papers published from 1947 onwards, Ernest Cheesman showed that Linnaeus' Musa sapientum and Musa paradisiaca were actually cultivars and descendants of two wild seed-producing species, Musa acuminata and Musa balbisiana, both first described by Luigi Aloysius Colla. He recommended the abolition of Linnaeus's species in favor of reclassifying bananas according to three morphologically distinct groups of cultivars – those primarily exhibiting the botanical characteristics of Musa balbisiana, those primarily exhibiting the botanical characteristics of Musa acuminata, and those with characteristics that are the combination of the two. Researchers Norman Simmonds and Ken Shepherd proposed a genome-based nomenclature system in 1955. This system eliminated almost all the difficulties and inconsistencies of the earlier classification of bananas based on assigning scientific names to cultivated varieties. Despite this, the original names are still recognized by some authorities today, leading to confusion.

The currently accepted scientific names for most groups of cultivated bananas are Musa acuminata Colla and Musa balbisiana Colla for the ancestral species, and Musa × paradisiaca L. for the hybrid M. acuminata × M. balbisiana.

Synonyms of M. × paradisica include:

- A large number of subspecific and varietial names of M. × paradisiaca, including M. p. subsp. sapientum (L.) Kuntze
- Musa × dacca Horan.
- Musa × sapidisiaca K.C.Jacob, nom. superfl.
- Musa × sapientum L., and a large number of its varietal names, including M. × sapientum var. paradisiaca (L.) Baker, nom. illeg.

Generally, modern classifications of banana cultivars follow Simmonds and Shepherd's system. Cultivars are placed in groups based on the number of chromosomes they have and which species they are derived from. Thus the Latundan banana is placed in the AAB Group, showing that it is a triploid derived from both M. acuminata (A) and M. balbisiana (B). For a list of the cultivars classified under this system see List of banana cultivars.

In 2012, a team of scientists announced they had achieved a draft sequence of the genome of Musa acuminata.

Bananas and Plantains

In regions such as North America and Europe, Musa fruits offered for sale can be divided into "bananas" and "plantains", based on their intended use as food. Thus the banana producer and distributor Chiquita produces publicity material for the American market which says that "a plantain is not a banana". The stated differences are that plantains are more starchy and less sweet; they are eaten cooked rather than raw; they have thicker skin, which may be green, yellow or black; and they can be used at any stage of ripeness. Linnaeus made the same distinction between plantains and bananas when first naming two "species" of Musa. Members of the "plantain subgroup" of banana cultivars, most important as food in West Africa and Latin America, correspond to the Chiquita description, having long pointed fruit. They are described by Ploetz et al. as "true" plantains, distinct from other cooking bananas. The cooking bananas of East Africa belong to a different group, the East African Highland bananas, so would not qualify as "true" plantains on this definition.

An alternative approach divides bananas into dessert bananas and cooking bananas, with plantains being one of the subgroups of cooking bananas. Triploid cultivars derived solely from M. acuminata are examples of "dessert bananas", whereas triploid cultivars derived from the hybrid between M. acuminata and M. balbinosa (in particular the plantain subgroup of the AAB Group) are "plantains". Small farmers in Colombia grow a much wider range of cultivars than large commercial plantations. A study of these cultivars showed that they could be placed into at least three groups based on their characteristics: dessert bananas, non-plantain cooking bananas, and plantains, although there were overlaps between dessert and cooking bananas.

In Southeast Asia – the center of diversity for bananas, both wild and cultivated – the distinction between "bananas" and "plantains" does not work, according to Valmayor et al. Many bananas are used both raw and cooked. There are starchy cooking bananas which are smaller than those eaten raw. The range of colors, sizes and shapes is far wider than in those grown or sold in Africa, Europe or the Americas. Southeast Asian languages do not make the distinction between "bananas" and "plantains" that is made in English (and Spanish). Thus both Cavendish cultivars, the classic yellow dessert bananas, and Saba cultivars, used mainly for cooking, are called pisang in Malaysia and Indonesia, kluai in Thailand and chuoi in Vietnam. Fe'i bananas, grown and eaten in the islands of the Pacific, are derived from entirely different wild species than traditional bananas and plantains. Most Fe'i bananas are cooked, but Karat bananas, which are short and squat with bright red skins, very different from the usual yellow dessert bananas, are eaten raw.

In summary, in commerce in Europe and the Americas (although not in small-scale cultivation), it is possible to distinguish between "bananas", which are eaten raw, and "plantains", which are cooked. In other regions of the world, particularly India, Southeast Asia and the islands of the Pacific, there are many more kinds of banana and the two-fold distinction is not useful and not made in local languages. Plantains are one of many kinds of cooking bananas, which are not always distinct from dessert bananas.

Historical cultivation

Early Cultivation

Farmers in Southeast Asia and Papua New Guinea first domesticated bananas. Recent archaeological and palaeoenvironmental evidence at Kuk Swamp in the Western Highlands Province of Papua New Guinea suggests that banana cultivation there goes back to at least 5000 BCE, and possibly to 8000 BCE. It is likely that other species were later and independently domesticated elsewhere in Southeast Asia. Southeast Asia is the region of primary diversity of the banana. Areas of secondary diversity are found in Africa, indicating a long history of banana cultivation in the region.Phytolith discoveries in Cameroon dating to the first millennium BCE triggered an as yet unresolved debate about the date of first cultivation in Africa. There is linguistic evidence that bananas were known in Madagascar around that time. The earliest prior evidence indicates that cultivation dates to no earlier than late 6th century CE. It is likely, however, that bananas were brought at least to Madagascar if not to the East African coast during the phase of Malagasy colonization of the island from South East Asia c. 400 CE.

The banana may also have been present in isolated locations elsewhere in the Middle East on the eve of Islam. The spread of Islam was followed by far-reaching diffusion. There are numerous references to it in Islamic texts (such as poems and hadiths) beginning in the 9th century. By the 10th century the banana appears in texts from Palestine and Egypt. From there it diffused into North Africa and Muslim Iberia. During the medieval ages, bananas from Granada were considered among the best in the Arab world. In 650, Islamic conquerors brought the banana to Palestine. Today, banana consumption increases significantly in Islamic countries during Ramadan, the month of daylight fasting.

Bananas were certainly grown in the Christian Kingdom of Cyprus by the late medieval period. Writing in 1458, the Italian traveller and writer Gabriele Capodilista wrote favourably of the extensive farm produce of the estates at Episkopi, near modern-day Limassol, including the region's banana plantations.

Bananas were introduced to the Americas by Portuguese sailors who brought the fruits from West Africa in the 16th century.

Many wild banana species as well as cultivars exist in extraordinary diversity in India, China, and Southeast Asia.

There are fuzzy bananas whose skins are bubblegum pink; green-and-white striped bananas with pulp the color of orange sherbet; bananas that, when cooked, taste like strawberries. The Double Mahoi plant can produce two bunches at once. The Chinese name of the aromatic Go San Heong banana means 'You can smell it from the next mountain.' The fingers on one banana plant grow fused; another produces bunches of a thousand fingers, each only an inch long.

— Mike Peed, The New Yorker

In 1999 archaeologists in London discovered what they believed to be the oldest banana in the UK, in a Tudor rubbish tip.

Plantation cultivation in the Caribbean, Central and South America

In the 15th and 16th centuries, Portuguese colonists started banana plantations in the Atlantic Islands, Brazil, and western Africa. North Americans began consuming bananas on a small scale at very high prices shortly after the Civil War, though it was only in the 1880s that it became more widespread. As late as the Victorian Era, bananas were not widely known in Europe, although they were available.Jules Verne introduces bananas to his readers with detailed descriptions in Around the World in Eighty Days (1872).

The earliest modern plantations originated in Jamaica and the related Western Caribbean Zone, including most of Central America. It involved the combination of modern transportation networks of steamships and railroads with the development of refrigeration that allowed bananas to have more time between harvesting and ripening. North America shippers like Lorenzo Dow Baker and Andrew Preston, the founders of the Boston Fruit Company started this process in the 1870s, but railroad builders like Minor C Keith also participated, eventually culminating in the multi-national giant corporations like today's Chiquita Brands International and Dole. These companies were monopolistic, vertically integrated (meaning they controlled growing, processing, shipping and marketing) and usually used political manipulation to build enclave economies (economies that were internally self-sufficient, virtually tax exempt, and export oriented that contribute very little to the host economy). Their political maneuvers, which gave rise to the term Banana republic for states like Honduras and Guatemala, included working with local elites and their rivalries to influence politics or playing the international interests of the United States, especially during the Cold War, to keep the political climate favorable to their interests.

Peasant Cultivation for Export in the Caribbean

The vast majority of the world's bananas today are cultivated for family consumption or for sale on local markets. India is the world leader in this sort of production, but many other Asian and African countries where climate and soil conditions allow cultivation also host large populations of banana growers who sell at least some of their crop.

There are peasant sector banana growers who produce for the world market in the Caribbean, however. The Windward Islands are notable for the growing, largely of Cavendish bananas, for an international market, generally in Europe but also in North America. In the Caribbean, and especially in Dominica where this sort of cultivation is widespread, holdings are in the 1–2 acre range. In many cases the farmer earns additional money from other crops, from engaging in labor outside the farm, and from a share of the earnings of relatives living overseas. This style of cultivation often was popular in the islands as bananas required little labor input and brought welcome extra income. Banana crops are vulnerable to destruction by high winds, such as tropical storms or cyclones.

After the signing of the NAFTA agreements in the 1990s, however, the tide turned against peasant producers. Their costs of production were relatively high and the ending of favorable tariff and other supports, especially in the European Economic Community, made it difficult for peasant producers to compete with the bananas grown on large plantations by the well capitalized firms like Chiquita and Dole. Not only did the large companies have access to cheap labor in the areas they worked, but they were better able to afford modern agronomic advances such as fertilization. The "dollar banana" produced by these concerns made the profit margins for peasant bananas unsustainable.

Caribbean countries have sought to redress this problem by providing government supported agronomic services and helping to organize producers' cooperatives. They have also been supporters of the Fair Trade movement which seeks to balance the inequities in the world trade in commodities.

East Africa

Most farms supply local consumption. Cooking bananas represent a major food source and a major income source for smallhold farmers. In east Africa, highland bananas are of greatest importance as a staple food crop. In countries such as Uganda, Burundi, and Rwanda per capita consumption has been estimated at 45 kilograms (99 lb) per year, the highest in the world.

Modern Cultivation

All widely cultivated bananas today descend from the two wild bananas Musa acuminata and Musa balbisiana. While the original wild bananas contained large seeds, diploid or polyploid cultivars (some being hybrids) with tiny seeds are preferred for human raw fruit consumption. These are propagated asexually from offshoots. The plant is allowed to produce two shoots at a time; a larger one for immediate fruiting and a smaller "sucker" or "follower" to produce fruit in 6–8 months. The life of a banana plantation is 25 years or longer, during which time the individual stools or planting sites may move slightly from their original positions as lateral rhizome formation dictates.

Cultivated bananas are parthenocarpic, i.e. the flesh of the fruit swells and ripens without its seeds being fertilized and developing. Lacking viable seeds, propagation typically involves farmers removing and transplanting part of the underground stem (called a corm). Usually this is done by carefully removing a sucker (a vertical shoot that develops from the base of the banana pseudostem) with some roots intact. However, small sympodial corms, representing not yet elongated suckers, are easier to transplant and can be left out of the ground for up to two weeks; they require minimal care and can be shipped in bulk.

It is not necessary to include the corm or root structure to propagate bananas; severed suckers without root material can be propagated in damp sand, although this takes somewhat longer.

In some countries, commercial propagation occurs by means of tissue culture. This method is preferred since it ensures disease-free planting material. When using vegetative parts such as suckers for propagation, there is a risk of transmitting diseases (especially the devastating Panama disease).

As a non-seasonal crop, bananas are available fresh year-round.

Cavendish

In global commerce in 2009, by far the most important cultivars belonged to the triploid AAA group of Musa acuminata, commonly referred to as Cavendish group bananas. They accounted for the majority of banana exports, despite only coming into existence in 1836.The cultivars Dwarf Cavendish and Grand Nain (Chiquita Banana) gained popularity in the 1950s after the previous mass-produced cultivar, Gros Michel (also an AAA group cultivar), became commercially unviable due to Panama disease, caused by the fungus Fusarium oxysporum which attacks the roots of the banana plant. Cavendish cultivars are resistant to the Panama Disease but in 2013 there were fears that the Black Sigatoka fungus would in turn make Cavendish bananas unviable.

Ease of transport and shelf life rather than superior taste make the Dwarf Cavendish the main export banana.

Even though it is no longer viable for large scale cultivation, Gros Michel is not extinct and is still grown in areas where Panama disease is not found. Likewise, Dwarf Cavendish and Grand Nain are in no danger of extinction, but they may leave supermarket shelves if disease makes it impossible to supply the global market. It is unclear if any existing cultivar can replace Cavendish bananas, so various hybridisation and genetic engineering programs are attempting to create a disease-resistant, mass-market banana.

Ripening

Export bananas are picked green, and ripen in special rooms upon arrival in the destination country. These rooms are air-tight and filled with ethylene gas to induce ripening. The vivid yellow color consumers normally associate with supermarket bananas is, in fact, caused by the artificial ripening process. Flavor and texture are also affected by ripening temperature. Bananas are refrigerated to between 13.5 and 15 °C (56.3 and 59.0 °F) during transport. At lower temperatures, ripening permanently stalls, and the bananas turn gray as cell walls break down. The skin of ripe bananas quickly blackens in the 4 °C (39 °F) environment of a domestic refrigerator, although the fruit inside remains unaffected.

"Tree-ripened" Cavendish bananas have a greenish-yellow appearance which changes to a brownish-yellow as they ripen further. Although both flavor and texture of tree-ripened bananas is generally regarded as superior to any type of green-picked fruit,this reduces shelf life to only 7–10 days.

Bananas can be ordered by the retailer "ungassed" (i.e. not treated with ethylene), and may show up at the supermarket fully green. Guineos verdes (green bananas) that have not been gassed will never fully ripen before becoming rotten. Instead of fresh eating, these bananas are best suited to cooking, as seen in Mexican culinary dishes.

A 2008 study reported that ripe bananas fluoresce when exposed to ultraviolet light. This property is attributed to the degradation of chlorophyll leading to the accumulation of a fluorescent product in the skin of the fruit. The chlorophyll breakdown product is stabilized by a propionate ester group. Banana-plant leaves also fluoresce in the same way. Green bananas do not fluoresce. The study suggested that this allows animals which can see light in the ultraviolet spectrum (tetrachromats and pentachromats) to more easily detect ripened bananas.

Storage and Transport

Bananas must be transported over long distances from the tropics to world markets. To obtain maximum shelf life, harvest comes before the fruit is mature. The fruit requires careful handling, rapid transport to ports, cooling, and refrigerated shipping. The goal is to prevent the bananas from producing their natural ripening agent, ethylene. This technology allows storage and transport for 3–4 weeks at 13 °C (55 °F). On arrival, bananas are held at about 17 °C (63 °F) and treated with a low concentration of ethylene. After a few days, the fruit begins to ripen and is distributed for final sale. Unripe bananas can not be held in home refrigerators because they suffer from the cold.Ripe bananas can be held for a few days at home. If bananas are too green, they can be put in a brown paper bag with an apple or tomato overnight to speed up the ripening process.

Carbon dioxide (which bananas produce) and ethylene absorbents extend fruit life even at high temperatures. This effect can be exploited by packing banana in a polyethylene bag and including an ethylene absorbent, e.g., potassium permanganate, on an inert carrier. The bag is then sealed with a band or string. This treatment has been shown to more than double lifespans up to 3–4 weeks without the need for refrigeration.

Production and export

Statistics on the production and export of bananas and plantains are available from the Food and Agriculture Organization. Some countries produce statistics which distinguish between bananas and plantains, but two of the top three producers (India and China) do not, so comparisons can only be made using the total for bananas and plantains combined. The 2012 statistics show that India led the world in banana production, producing around 18% of the worldwide crop of 139 million metric tonnes. Philippines was the next largest producer with around 7% of the worldwide crop. Its national data does distinguish between bananas and plantains, and shows that the latter made up over 95% of production. Ten countries produced around two thirds of the total world production.

The statistics for the export of bananas and plantains show a rather different picture. Total world exports at around 18 million metric tonnes amounted to only 12% of total world production; two thirds of the exports were generated by only five countries. The top three producing countries do not appear in this table, and two countries, Costa Rica and Guatemala, do not appear in the table of top producers. Only the Philippines has a consistent position in both tables. Exports were dominated by Ecuador, with 29% of the world total. Statistics for Ecuador distinguish between bananas and plantains; 93% of its exports were classified as bananas.

Bananas and plantains constitute a major staple food crop for millions of people in developing countries. In most tropical countries, green (unripe) bananas used for cooking represent the main cultivars. Bananas are cooked in ways that are similar to potatoes. Both can be fried, boiled, baked, or chipped and have similar taste and texture when served. One banana provides about the same calories as one potato.

Most producers are small-scale farmers either for home consumption or local markets. Because bananas and plantains produce fruit year-round, they provide an extremely valuable food source during the hunger season (when the food from one annual/semi-annual harvest has been consumed, and the next is still to come). Bananas and plantains are therefore critical to global food security.

Bananas have been an important source of disagreement in the Doha Round of trade talks. A study for ICTSD showed that the new deal on EU banana import tariffs will be a boon to Latin American exporters but would trigger a drop in exports of the fruit from African, Caribbean and Pacific (ACP) countries.

Bananas are among the most widely consumed foods in the world. Chiquita, Del Monte, Dole, and Fyffes grow their own bananas in Ecuador, Colombia, Costa Rica, Guatemala, and Honduras. Banana plantations are capital intensive and demand significant expertise. The majority of independent growers are large and wealthy landowners in these countries. Producers have attempted to raise prices via marketing them as "fair trade" or Rainforest Alliance-certified in some countries.

The banana has an extensive trade history starting with firms such as the Irish Fyffes and the US United Fruit Company (now Chiquita) at the end of the 19th century. For much of the 20th century, bananas and coffee dominated the export economies of Central America. In the 1930s, bananas and coffee made up as much as 75% of the region's exports. As late as 1960, the two crops accounted for 67% of the exports from the region. Though the two were grown in similar regions, they tended not to be distributed together. The United Fruit Company based its business almost entirely on the banana trade, because the coffee trade proved too difficult to control. The term "banana republic" has been applied to most countries in Central America, but from a strict economic perspective only Costa Rica, Honduras, and Panama had economies dominated by the banana trade.

The European Union has traditionally imported many of its bananas from former European Caribbean colonies, paying guaranteed prices above global market rates. These arrangements have now been largely withdrawn under pressure from other major trading powers, principally the United States. The withdrawal of these indirect subsidies to Caribbean producers is expected to favour the more efficient banana producers of Central America, in which American companies have an economic interest. In addition, small-scale Caribbean producers are finding it difficult to comply with increasingly strict certification requirements. Some support is being provided to Caribbean countries under the EU's Banana Accompanying Measures (BAM).

The United States produces few bananas. A mere 14,000 tonnes (14,000 long tons; 15,000 short tons) were grown in Hawaii in 2001. Bananas were once grown in Florida and southern California.

In March 2014 it was announced that Fyffes and Chiquita would merge to create the world's largest banana company, worth about $1bn (£597m). The new firm, named ChiquitaFyffes, is expected to sell about 160 million boxes of bananas annually.

Pests, diseases, and natural disasters

While in no danger of outright extinction, the most common edible banana cultivar Cavendish (extremely popular in Europe and the Americas) could become unviable for large-scale cultivation in the next 10–20 years. Its predecessor 'Gros Michel', discovered in the 1820s, suffered this fate. Like almost all bananas, Cavendish lacks genetic diversity, which makes it vulnerable to diseases, threatening both commercial cultivation and small-scale subsistence farming. Some commentators remarked that those variants which could replace what much of the world considers a "typical banana" are so different that most people would not consider them the same fruit, and blame the decline of the banana on monogenetic cultivation driven by short-term commercial motives.

Panama Disease

Panama disease is caused by a fusarium soil fungus (Race 1), which enters the plants through the roots and travels with water into the trunk and leaves, producing gels and gums that cut off the flow of water and nutrients, causing the plant to wilt, and exposing the rest of the plant to lethal amounts of sunlight. Prior to 1960, almost all commercial banana production centered on "Gros Michel", which was highly susceptible.Cavendish was chosen as the replacement for Gros Michel because, among resistant cultivars, it produces the highest quality fruit. However, more care is required for shipping the Cavendish, and its quality compared to Gros Michel is debated.

According to current sources, a deadly form of Panama disease is infecting Cavendish. All plants are genetically identical, which prevents evolution of disease resistance. Researchers are examining hundreds of wild varieties for resistance.

Tropical race

Tropical race 4 (TR4) is a reinvigorated strain of Panama disease first discovered in 1993. This virulent form of fusarium wilt has wiped out Cavendish in several southeast Asian countries. It has yet to reach the Americas; however, soil fungi can easily be carried on boots, clothing, or tools. This is how TR4 travels and is its most likely route into Latin America. Cavendish is highly susceptible to TR4, and over time, Cavendish is almost certain to be eliminated from commercial production by this disease. The only known defense to TR4 is genetic resistance.

Black sigatoka

Black sigatoka is a fungal leaf spot disease first observed in Fiji in 1963 or 1964. Black Sigatoka (also known as black leaf streak) has spread to banana plantations throughout the tropics from infected banana leaves that were used as packing material. It affects all main cultivars of bananas and plantains (including the Cavendish cultivars), impeding photosynthesis by blackening parts of the leaves, eventually killing the entire leaf. Starved for energy, fruit production falls by 50% or more, and the bananas that do grow ripen prematurely, making them unsuitable for export. The fungus has shown ever-increasing resistance to treatment, with the current expense for treating 1 hectare (2.5 acres) exceeding $1,000 per year. In addition to the expense, there is the question of how long intensive spraying can be environmentally justified. Several resistant cultivars of banana have been developed, but none has yet received commercial acceptance due to taste and texture issues.

In East Africa

With the arrival of black sigatoka, banana production in eastern Africa fell by over 40%. For example, during the 1970s, Uganda produced 15 to 20 tonnes (15 to 20 long tons; 17 to 22 short tons) of bananas per hectare. Today, production has fallen to only 6 tonnes (5.9 long tons; 6.6 short tons) per hectare.

The situation has started to improve as new disease-resistant cultivars have been developed by the International Institute of Tropical Agriculture and the National Agricultural Research Organisation of Uganda (NARO), such as FHIA-17 (known in Uganda as the Kabana 3). These new cultivars taste different from the Cabana banana, which has slowed their acceptance by local farmers. However, by adding mulch and manure to the soil around the base of the plant, these new cultivars have substantially increased yields in the areas where they have been tried.

The International Institute of Tropical Agriculture and NARO, funded by the Rockefeller Foundation and CGIAR have started trials for genetically modified bananas that are resistant to both Black sigatoka and banana weevils. It is developing cultivars specifically for smallholder and subsistence farmers.

Banana bunchy top virus

Banana bunchy top virus (BBTV) jumps from plant to plant using aphids. It stunts leaves, resulting in a "bunched" appearance. Generally, an infected plant does not produce fruit, although mild strains exist which allow some production. These mild strains are often mistaken for malnourishment, or a disease other than BBTV. There is no cure; however, its effect can be minimized by planting only tissue-cultured plants (in vitro propagation), controlling aphids, and immediately removing and destroying infected plants.

Banana bacterial wilt

Banana bacterial wilt (BBW) is a bacterial disease caused by Xanthomonas campestris pv. musacearum. After being originally identified on a close relative of bananas, Ensete ventricosum, in Ethiopia in the 1960s, BBW occurred in Uganda in 2001 affecting all banana cultivars. Since then BBW has been diagnosed in Central and East Africa including the banana growing regions of Rwanda, the Democratic Republic of the Congo, Tanzania, Kenya, Burundi, and Uganda.

Papaya

The papaya (from Carib via Spanish), papaw, or pawpaw is the plant Carica papaya, one of the 22 accepted species in the genus Carica of the family Caricaceae.

It is native to the tropics of the Americas, perhaps from southern Mexico and neighboring Central America. It was first cultivated in Mexico several centuries before the emergence of the Mesoamerican classical civilizations.

The papaya is a large, tree-like plant, with a single stem growing from 5 to 10 m (16 to 33 ft) tall, with spirally arranged leaves confined to the top of the trunk. The lower trunk is conspicuously scarred where leaves and fruit were borne. The leaves are large, 50–70 cm (20–28 in) in diameter, deeply palmately lobed, with seven lobes. Unusually for such large plants, the trees are dioecious. The tree is usually unbranched, unless lopped. The flowers are similar in shape to the flowers of the Plumeria, but are much smaller and wax-like. They appear on the axils of the leaves, maturing into large fruit - 15–45 cm (5.9–17.7 in) long and 10–30 cm (3.9–11.8 in) in diameter. The fruit is a type of berry. It is ripe when it feels soft (as soft as a ripe avocado or a bit softer) and its skin has attained an amber to orange hue.

Carica papaya was the first transgenic fruit tree to have its genome sequenced.

Cultivation and production

Papaya plants grow in three sexes: male, female, hermaphrodite. The male produces only pollen, never fruit. The female will produce small, inedible fruits unless pollinated.

The hermaphrodite can self-pollinate since its flowers contain both male stamens and female ovaries. Almost all commercial papaya orchards contain only hermaphrodites.

Originally from southern Mexico (particularly Chiapas and Veracruz), Central America, and northern South America, the papaya is now cultivated in most tropical countries. In cultivation, it grows rapidly, fruiting within three years. It is, however, highly frost-sensitive, limiting its production to tropical climates. Temperatures below "2 °C (29 °F) are greatly harmful if not fatal. In Florida and California, growth is generally limited to southern parts of the states. It prefers sandy, well-drained soil, as standing water will kill the plant within 24 hours.

For cultivation, however, only female plants are used, since they give off a single flower each time, and close to the base of the plant, while the male gives off multiple flowers in long stems, which result in poorer quality fruit.

India and Brazil are the major producers of papaya, together providing 57% of the world total of 12.4 million tons in 2013 (FAOSTAT chart).

Gaining in popularity among tropical fruits worldwide, papaya is now ranked fourth in total tropical fruit production after bananas, oranges, and mango. Global papaya production has grown significantly over the last few years, mainly as a result of increased production in India. Papaya has become an important agricultural export for developing countries, where export revenues of the fruit provide a livelihood for thousands of people, especially in Asia and Latin America.

Diseases and pests

Viruses

Papaya ringspot virus is a well-known virus within plants in Florida. The first signs of the virus are yellowing and vein-clearing of younger leaves, as well as mottling yellow leaves. Infected leaves may obtain blisters, roughen or narrow, with blades sticking upwards from the middle of the leaves. The petioles and stems may develop dark green greasy streaks and in time become shorter. The ringspots are circular, C-shaped markings that are darker green than the fruit itself. In the later stages of the virus, the markings may become gray and crusty. Viral infections impact growth and reduce the fruit's quality. One of the biggest effects that viral infections have on papaya is the taste. As of 2010, the only way to protect papaya from this virus is genetic modification.

The papaya mosaic virus destroys the plant until only a small tuft of leaves are left. The virus affects both the leaves of the plant and the fruit. Leaves show thin, irregular, dark-green lines around the borders and clear areas around the veins. The more severely affected leaves are irregular and linear in shape. The virus can infect the fruit at any stage of its maturity. Fruits as young as 2 weeks old have been spotted with dark-green ringspots about 1 inch in diameter. Rings on the fruit are most likely seen on either the stem end or the blossom end. In the early stages of the ringspots, the rings tend to be many closed circles, but as the disease develops, the rings will increase in diameter consisting of one large ring. The difference between the ringspot and the mosaic viruses is the ripe fruit in the ringspot has mottling of colors and mosaic does not.

Fungi

The fungus anthracnose is known to specifically attack papaya, especially the mature fruits. The disease starts out small with very few signs, such as water-soaked spots on ripening fruits. The spots become sunken, turn brown or black, and may get bigger. In some of the older spots, the fungus may produce pink spores. The fruit ends up being soft and having an off flavor because the fungus grows into the fruit.

The fungus powdery mildew occurs as a superficial white presence on the surface of the leaf in which it is easily recognized. Tiny, light yellow spots begin on the lower surfaces of the leaf as the disease starts to make its way. The spots enlarge and white powdery growth appears on the leaves. The infection usually appears at the upper leaf surface as white fungal growth. Powdery mildew is not as severe as other diseases.

The fungus phytophthora blight causes damping-off, root rot, stem rot, stem girdling, and fruit rot. Damping-off happens in young plants by wilting and death. The spots on established plants start out as white, water-soaked lesions at the fruit and branch scars. These spots enlarge and eventually cause death. The most dangerous feature of the disease is the infection of the fruit which may be toxic to consumers. The roots can also be severely and rapidly infected, causing the plant to brown and wilt away, collapsing within days.

Pests

The papaya fruit fly is mainly yellow with black marks. The female papaya fruit fly has a very long, slender abdomen with an extended ovipositor that exceeds the length of its body. The male papaya fruit fly looks like the female with the differences of a hairy abdomen and no ovipositor. Long, slender eggs are laid inside of the fruit by the female papaya fruit fly. The larvae are white and look very much like the regular fruit fly larvae. The female is capable of laying 100 or more eggs, and they are laid during the evening or early morning in groups of ten inside young fruit. They usually hatch within 12 days of being in the fruit where they feed on the seeds and interior parts of the fruit. When the larvae mature (usually 16 days after being hatched) they eat their way out of the fruit, drop to the ground, and pupate just below the soil and emerge within one to two weeks as mature flies. The flesh of the papaya must be ripe for the fly to migrate towards the surface of the fruit because unripe papaya juice is fatal to them. The papaya will turn yellow and drop to the ground if it is infected by the papaya fruit fly.

The two-spotted spider mite is a 0.5-mm-long brown or orange-red or a green, greenish yellow translucent oval pest. They all have needle-like piercing-sucking mouthparts and feed by piercing the plant tissue with their mouthparts, usually on the underside of the plant. The spider mites spin fine threads of webbing on the host plant, and when they remove the sap, the mesophyll tissue collapses and a small chlorotic spot forms at the feeding sites. The leaves of the papaya fruit turn yellow, gray, or bronze. If the spider mites are not controlled, they can cause the death of the fruit.

The papaya whitefly lays yellow, oval eggs that appear dusted on the undersides of the leaves. They eat papaya leaves, therefore damaging the fruit. There, the eggs developed into flies in three stages called instars. The first instar has well-developed legs and is the only mobile immature life stage. The crawlers insert their mouthparts in the lower surfaces of the leaf when they find it suitable and usually do not move again in this stage. The next instars are flattened, oval, and scale-like. In the final stage, the pupal whiteflies are more convex, with large, conspicuously red eyes.

Cultivars

Two kinds of papayas are commonly grown. One has sweet, red or orange flesh, and the other has yellow flesh; in Australia, these are called "red papaya" and "yellow papaw", respectively. Either kind, picked green, is called a "green papaya".

The large-fruited, red-fleshed 'Maradol', 'Sunrise', and 'Caribbean Red' papayas often sold in U.S. markets are commonly grown in Mexico and Belize.

In 2011, Philippine researchers reported that by hybridizing papaya with Vasconcellea quercifolia, they had developed conventionally bred, nongenetically engineered papaya resistant to PRV.

Genetically engineered cultivars

In response to the papaya ringspot virus (PRV) outbreak in Hawaii, in 1998, genetically altered papaya were approved and brought to market (including 'SunUp' and 'Rainbow' varieties.) Varieties resistant to PRV have some DNA of this virus incorporated into the DNA of the plant. As of 2010, 80% of Hawaiian papaya plants were genetically modified. The modifications were made by University of Hawaii scientists who made the modified seeds available to farmers without charge.

Culinary uses

The ripe fruit of the papaya is usually eaten raw, without skin or seeds. The unripe green fruit can be eaten cooked, usually in curries, salads, and stews. Green papaya is used in Southeast Asian cooking, both raw and cooked. In Thai cuisine, papaya is used to make Thai salads such as som tam and Thai curries such as kaeng som when still not fully ripe. In Indonesian cuisine, the unripe green fruits and young leaves are boiled for use as part of lalab salad, while the flower buds are sautéed and stir-fried with chillies and green tomatoes as Minahasan papaya flower vegetable dish. Papayas have a relatively high amount of pectin, which can be used to make jellies. The smell of ripe, fresh papaya flesh can strike some people as unpleasant. In Brazil, the unripe fruits are often used to make sweets or preserves.

The black seeds of the papaya are edible and have a sharp, spicy taste. They are sometimes ground and used as a substitute for black pepper.

In some parts of Asia, the young leaves of the papaya are steamed and eaten like spinach.

Meat tenderizing

Both green papaya fruit and the tree's latex are rich in papain, a protease used for tenderizing meat and other proteins, as practiced currently by indigenous Americans and people of the Caribbean region. It is now included as a component in some powdered meat tenderizers.

Phytochemicals

Papaya skin, pulp and seeds contain a variety of phytochemicals, including carotenoids and polyphenols, as well as benzyl isothiocyanates and benzyl glucosinates having skin and pulp levels that increase during ripening. Papaya seeds also contain the cyanogenic substance, prunasin.

Traditional medicine

In some parts of the world, papaya leaves are made into tea as a treatment for malaria, but the mechanism is not understood and no treatment method based on these results has been scientifically proven.

9

OILSEEDS: PRODUCTION AND MANAGEMENT

The important oil seeds are: Mustard (Brassica camprestris); Groundnut (Arachis hypogaea), Rapeseed (Brassica spp), Soybean (Glax mexican), Linseed, Till, Saf- flower (Carthomus tinctorius), Castor seed (Ricinus communis), Sunflower (Helianthus annus), Niger seed (Guizotia abyssinica).

Oilseed has drawn the attention of the government because much of the foreign exchange is lost in importing oilseeds from oilseed producing countries. Now, India is the third largest oilseed economy in the world after China. It occupies a distinct position not in terms of area under cultivation but also in terms of diversity in cultivated oilseeds.

The country is blessed with agro-ecological conditions favourable for the cultivation of nine oilseeds of which seven are edible. The edible oilseeds are : groundnut, rapeseed, mustard, soybean, sunflower, safflower, sesame, and niger, the non-edible are castor and linseed.

India is the larger producer of sesame and castor and second in the world in the production of groundnut, rapeseed, mustard next to China. In 1996-97 India produced 24 million tonnes accounting for 9.7 per cent of global oilseed production. Indian vegetable oil industry achieved a domestic turnover of Rs. 35,000 crores.

But there is no room for complacency because:

1. There has been import of 15 lakh tonnes as the demand would be of 26 million tonnes by the turn of the century.
2. There is a high degree of fluctuation in the annual production owing to cultivation predominately under low and uncertain rainfall and input starved conditions coupled with poor crop management.
3. There exists a wide commercially untapped yield reservoir in almost all oilseed crops.

4. There is a lot of potentials in oilseeds in terms of production, processing and export.

There are constraints in increasing productivity of oilseeds:

1. Biotic and Abiotic.
2. On farm and Off farm.
3. Technological and non-technological (institutional, Cultural, socio-economic and political).

The oilseed crops are subject to vagaries of nature (monsoon) resulting in lower yields as compared to irrigated crops. This situation demands the necessity of protective irrigation at critical stages. There is lack of infrastructure and continuous raising of crop on the same piece of land results in depletion of soil fertility and incidence of pest and diseases increase.

The progress in the introducing of HYV is Very poor particularly with built in resistance to abiotic and biotic stress. The low domestic prices compel the farmers to divert land to more paying crops. There is gap due to poor yield and also non-expansion of area under oilseeds which need to be filled.

The finding from a continuous research from 1988-89 to 1995-96 reveal the fact that yield could be raised to the extend of 25 to 152 per cent as there exists a wide commercially exploitable yield reservoir which will be more with technology improvement.

Since 1950-51, the oilseed, area, production and productivity have increased nearly 2.5 times, 5 times and 2 times respectively. But between 1993-97 there has been only marginal increase in vegetable oil prices in the domestic markets with the result, oilseed production nearly stagnated as farmers never accepted a combination of low prices and high risk.

Production and Technology of Oilseeds:

The annual vegetable oilseeds are cultivated in fourteen States of which eight, namely, M.P., Rajasthan, A.P., Gujarat, Karnataka, Maharashtra, U.P. and Tamil Nadu account for ninety per cent of the oilseeds area and production in the country. M.P., shares the highest area and production followed by Rajasthan, Andhra Pradesh, Gujarat.

Amongst the different oilseeds rapeseed and mustard happens to be the major oilseed crop in seven States, groundnut in four States, Soybean, sasamum, and niger in one State each.

At present, oil palm is cultivated on eight lakh hectares, as against the potential area of 0.8 million hectares, in the country spread across eleven States. With a productivity of nearly four tonnes a hectare oil palm can constitute about one million tonnes of edible oil in the near future accounting for about ten per cent of the demand.

Government formed the integrated policy on oilseed production which brought:

1. Self-sufficiency in vegetable oil.
2. Reduced instability in the production of oilseeds.
3. Massive savings in foreign exchange.
4. Clear policy management.
5. Smooth management of consequences of high prices, and
6. Movement towards modernization of oilseeds sector.

In the last two decades over 240 improved varieties/hybrids have been developed in annual oilseeds, which have shown 9-38 per cent yield superiority over their locally cultivated counterparts. Still there is scope for 50 per cent increase in output.

There are some HYV oilseed crops such as Sunflower hybrid DSH-129, developed by the Directorate of Oilseed Research Hyderabad released for commercial cultivation with a potential of 1750 kgs / hectare on an average. It is a matter of great pride.

During the rabi-summer season in black soils, the hybrid offer 22 per cent higher seed yields over the present varieties. Castor hybrids are: DCH-32 and GCH-4 and a high yielding variety DCS-9 (Jyoti) has revolutionized its production in the country.

There is an effort on breeding for increasing oil content in the seed, resistance to biotic and abiotic stresses and reducing crop duration, have been quite successful.

To maximize yield gains from improved varieties/hybrids, agronomic package for efficient crop management have been developed for different situations. Genetic resources in different oilseeds have been augmented through the collection of native and exotic plant species.

The National Bureau of Plant Genetics Resources, New Delhi, now maintains over 18,000 collections in oil crops which used to be exploited through conventional novel breeding approaches for developing high yielding varieties possessing resistance to biotic and abiotic stresses.

Integrated Nutrient Management of Oilseeds

Sustainability in production can come by integrated nutrient management— chemical fertilizer should be partially replaced by organic, green manures and bio-fertilizers i.e., rhizobium. Biological and industrial wastes should be used like cattle dung, crop residues, forest litters, rural composts, city refuges, sewage, sludge, press muds and industrial and domestic waste water fully utilized.

Green manuring crops such as dhaincha, sanai which are leguminous crops. These crops fix 49 to 152 kgs of atmospheric nitrogen per hectare. Where the crop is cultivated for the first time such as groundnut application of brady rhizobium to groundnut prove beneficial.

The brady rhizobium strains are: KGR-6 & IGR-40 tolerant to seed treating fungi Soybean be inoculated with Rhizobium japoniccum which is highly effective but the soil should not be deficient in phosphorus, calcium, and molybdenum. Where soil is poor in phosphorus phosphate solubilizing organisms like phospho-bacter and mycorrhriza have proved efficient to fix these organism.

Judicious use of Pesticides in Production of Oilseeds:

The problem associated with the use of pesticides has become alarming because mustard aphis and halithosis groundnut miners have become resistant, for this problem the answer lies in Integrated Pest Management. Against these pest use of resistant varieties must be used.

Marketing and Export of Oilseeds:

Technology Mission on Oilseeds:

A technology mission on oilseed was set up in 1986 to cut across all barriers in government, industry and the farming community to increase production of oilseeds and reduce the foreign exchange depletion due to import. Eventually, the mission succeeded in growing oilseeds and earning foreign exchange.

T.M.O. developed an integrated policy framework which harmonized the interest of the four partners in this endeavour on cost-effective basis:

1. For the farmers: An incentive price, profitable technology and efficient supply of inputs and services.

2. For the consumers: An affordable price for edible oils.

3. For the processing industry and trade: Plans of raw materials and better technology to cut down unit cost of processing and to set up export, and

4. For the economy: Faster growth with higher oilseed production, lower oil imports, and enhanced export of oilseed derivatives. The TMO worked at the national, state and field levels committees of farmers, their cooperative societies, agricultural research and extension scientists, input suppliers, banks, processing experts, and industry, traders and others to achieve the above goals.

Action plans were drawn up indicating the objectives, targets, constraints, strategies, resources, time schedules and accountability for every activity such as research, extension, input supply, procurement price support, storage, construction and processing efficiency.

Management sciences was harnessed not only to organize all supporting activities, but also to monitor them, identify weaknesses and active remedial measures without loss of time. An outside expert agency was hired for concurrent evaluation through interview with a statistically significant number of oilseed farmers. The agency reported how the project was progressing and identified the cause correction.

National Dairy Development Board's Entry: The National Dairy Development Board — the one reputed for White Revolution was called upon to engineer the oilseed (Yellow Revolution) which initiated an oilseed market intervention operation soon after harvest to help the farmers get a better share of the consumer's rupee.

Its 4,000 cooperative societies provided their one million farmer members incentive prices, production technology and good quality inputs. These societies combined together into unions and federations to set up their own processing units on modern lines, so that the farmers benefited from economics of efficient processing operations.

Oilseed farmers were motivated to organize exclusive cooperatives in nine States where oilseeds are important crop, Gujarat, Maharashtra, Tamil Nadu, Andhra Pradesh, Madhya Pradesh, Rajasthan, Uttar Pradesh, and Orissa.

A viable primary society for a group of villages would have around 250 growers with about 500 hectares; the society would have an office and a godown for storing inputs such as seeds, and fertilizers as well as oilseeds produced by members to be marketed by the society.

About 100 such primary societies would be able to pool their resources together to setup a cooperative union, a processing unit, with financial assistance from the NDDB could be setup for each union. Five or six unions could setup the State oilseed Federation in the nine selected States.

Professional agriculturists could be employed to provide the best technical support for cultivation by farmers and the union would setup State-of-the-Art processing and storage units to reap the economics of scale. Which would return to the member farmers as dividend.

Under the NDDB master plan (1985) to establish by 1993 Primary Societies— 6,000 in seven States with sixty-nine unions, and 1.3 million oilseed farmers as members, cultivating 2.5 million hectares of oilseed land.

Two more States U.P. and Rajasthan were added to it in 1992 thus 5,017 Societies established with 8.2 lakhs growers cultivating 19.24 lakh hectares.

Nine State Level Federations had been setup with a processing facilities of 3.3 tonnes per day oilseeds crushing 1980 tonnes per day solvent extraction and 733 tonnes per day refining capacity. In addition, a storage capacity of 1.7 lakh tonnes oilseeds and 2.77 lakh tonnes of oil has been created.

Significant progress has been made in the distribution of agricultural inputs and in the dissemination of International Crop Research Institute for semi-arid tropics groundnut production and dry-land farming technologies.

The nine federations have forged strong links with the extension wings of the ICAR as well as these nine States Agricultural Departments for training have been done. This has brought an overall healthy and fair deal for all concerned.

Export of Oilseeds and their Products:

Export from oilseeds witnessed a sharp recovery during 1995-96 after a set back during 1994-95. Export touched an all-time height Rs. 2,656 crores as against Rs. 1,867 crores in 1994-95 and the previous year high of Rs. 2,289 crores achieved during 1993-94. Commodities such as oil cakes, castor oil, HPS groundnut, sesame, seeds, and extraction of rapeseed, sunflower and groundnut are the main export items.

Policy Options for Oilseeds:

(a) Research by Government: Research should be supported by government funds till it is self-sufficient, research be prioritized. The support price is important as the crop is sensitive to market mechanism and must be based on quality (oil centre) consideration to encourage researchers and farmers.

To keep the imported prices of oil a little higher than the indigenous prices of oil it is necessary to adjust import duty in accordance to the peak and lean period keeping high and low duty respectively which is called "seasonally variable import Duty". Hence, integration of the interest of producers (farmers), processors and oilseed sector be brought for substantially and balanced growth.

Future trading in oilseeds is a welcome proposition for the stabilization of prices in the long run. The level of international future exchanges is also warranted in view of India's major share in the global output (63%) and trade (75%) at least in castor. The import of split soybean is a welcome proposition as it would bring development for the industry which is confronted with the problem of surplus capacity.

The policy of reserving crushing by small units has brought inefficiency. Phytosanitary measures in trade is needed to minimize residue of pesticides in products and by-products. Irrigation charges be in accordance to the use than a flat rate as irrigation consumption is less in oilseeds crops.

International linkage is needed for the overall development of the oilseed sector in terms of commercial exploitation of untapped reservoirs and value addition. Export need encouragement with surplus output viz., rapeseed, mustard and oilseed should be internationally competitive.

There is a very prospective use as a value added product of oilseeds, for instance, sesamum seeds and sunflower seed in confectionery, special erucic rich oil, plasticize additives and oleo chemicals from castor oil, automobile fuel substitute from rapeseed oil, cleavage product from fatty acids and fatty alcohols, short chain fatty acids for quick regenerative tissue growth in case of fire accidents and plastic surgery in human etc.

The above policies would definitely bring sustained high production and income to farmers, higher efficiency and profitability to processors and quality oil at affordable prices to consumers at large.

(b) Price Trends: A look at the table giving two consecutive years' prices i.e., 1988-89 and 1989-90 we see a uniform increase in prices of different oils which shows that the prices increase as a matter of fact with the inflationary tendencies but there is uniformity in the difference in their prices and its rise.

(c) Price Stabilization: Edible oil is a very sensitive commodity. Like other food articles edible oil prices have gone up and to an extent of six per cent on account of increased money supply. A buffer stock of ten per cent of the national output need to be maintained to keep prices within reasonable limits.

Thus, oilseed be purchased firm the farmers at an incentive prices. NDDB could not discharge the control. The output oilseed needs increase and modernization of processing industry.

Technical Mission Operation (TMO) need to be given full support in its endeavour to implement the integrated policy declared by government in 1988, as this policy can help step up production and exports and also reduce imports on cost-effective basis.

Peanut

The peanut, also known as the groundnut and the goober and taxonomically classified as Arachis hypogaea, is a legume crop grown mainly for its edible seeds. It is widely grown in the tropics and subtropics, being important to both small and large commercial producers. It is classified as both a grain legume and, because of its high oil content, an oil crop. World annual production of shelled peanuts was 42 million tonnes in 2014. Atypically among crop plants, peanut pods develop underground rather than aboveground. It is this characteristic that the botanist Linnaeus used to assign the specific name hypogaea, which means "under the earth."

As a legume, the peanut belongs to the botanical family Fabaceae; this is also known as Leguminosae, and commonly known as the bean, or pea, family. Like most other legumes, peanuts harbor symbiotic nitrogen-fixing bacteria in root nodules. This capacity to fix nitrogen means peanuts require less nitrogen-containing fertilizer and improve soil fertility, making them valuable in crop rotations.

Peanuts are similar in taste and nutritional profile to tree nuts such as walnuts and almonds, and are often served in similar ways in Western cuisines. The botanical definition of a "nut" is a fruit whose ovary wall becomes very hard at maturity. Using this criterion, the peanut is not a true nut, but rather a legume. However, for culinary purposes and in common English language usage, peanuts are usually referred to as nuts.

History

Cultivated peanut (A. hypogaea) arose from a hybrid between two wild species of peanut, thought to be A. duranensis and A. ipaensis. The initial hybrid would have been sterile, but spontaneous chromosome doubling restored its fertility, forming what is termed an amphidiploid or allotetraploid. Genetic analysis suggests the hybridization event probably occurred only once and gave rise to A. monticola, a wild form of peanut that occurs in a few restricted locations in northwestern Argentina, and by artificial selection to A. hypogaea. The process of domestication through artificial selection made A. hypogaea dramatically different from its wild relatives. The domesticated plants are more bushy and compact, and have a different pod structure and larger seeds. The initial domestication may have taken place in northwestern Argentina, or in southeastern Bolivia, where the peanut landraces with the most wild-like features are grown today. From this primary center of origin, cultivation spread and formed secondary and tertiary centers of diversity in Peru, Ecuador, Brazil, Paraguay, and Uruguay. Over time, thousands of peanut landraces evolved; these are classified into six botanical varieties and two subspecies (as listed in the peanut scientific classification table). Subspecies A. h. fastigiata types are more upright in their growth habit and have shorter crop cycles. Subspecies A. h. hypogaea types spread more on the ground and have longer crop cycles.

The oldest known archeological remains of pods have been dated at about 7,600 years old. These may be pods from a wild species that was in cultivation, or A. hypogaea in the early phase of domestication. They were found in Peru, where dry climatic conditions are favorable to the preservation of organic material. Almost certainly, peanut cultivation antedated this at the center of origin where the climate is moister. Many pre-Columbian cultures, such as the Moche, depicted peanuts in their art. Cultivation was well established in Mesoamerica before the Spanish arrived.

There, the conquistadors found the tlalcacahuatl (the plant's Nahuatl name, whence Mexican Spanish cacahuate, Castillian Spanish cacahuete, and French cacahuète) being offered for sale in the marketplace of Tenochtitlan. The peanut was later spread worldwide by European traders, and cultivation is now very widespread in tropical and subtropical regions. In West Africa, it substantially replaced a crop plant from the same family, the Bambara groundnut, whose seed pods also develop underground. In Asia, it became an agricultural mainstay and this region is now the largest producer in the world.

In the English-speaking world, peanut growing is most important in the United States. Although it was mainly a garden crop for much of the colonial period, it was mostly used as animal feed stock until the 1930s. The US Department of Agriculture initiated a program to encourage agricultural production and human consumption of peanuts in the late 19th and early 20th centuries. George Washington Carver developed hundreds of recipes for peanuts during his tenure in the program.

Cultivation

Peanuts grow best in light, sandy loam soil with a pH of 5.9–7. Their capacity to fix nitrogen means that, providing they nodulate properly, peanuts benefit little or not at all from nitrogen-containing fertilizer, and they improve soil fertility. Therefore, they are valuable in crop rotations. Also, the yield of the peanut crop itself is increased in rotations, through reduced diseases, pests and weeds. For instance, in Texas, peanuts in a three-year rotation with corn yield 50% more than nonrotated peanuts. Adequate levels of phosphorus, potassium, calcium, magnesium, and micronutrients are also necessary for good yields. To develop well, peanuts need warm weather throughout the growing season. They can be grown with as little as 350 mm of water, but for best yields need at least 500 mm. Depending on growing conditions and the cultivar of peanut, harvest is usually 90 to 130 days after planting for subspecies A. h. fastigiata types, and 120 to 150 days after planting for subspecies A. h. hypogaea types. Subspecies A. h. hypogaea types yield more, and are usually preferred where the growing seasons are long enough.

Peanut plants continue to produce flowers when pods are developing, therefore even when they are ready for harvest, some pods are immature. The timing of harvest is an important decision to maximize yield. If it is too early, too many pods will be unripe. If too late, the pods will snap off at the stalk, and will remain in the soil. For harvesting, the entire plant, including most of the roots, is removed from the soil. The pods are covered with a network of raised veins and are constricted between seeds.

Harvesting occurs in two stages: In mechanized systems, a machine is used to cut off the main root of the peanut plant by cutting through the soil just below the level of the peanut pods. The machine lifts the "bush" from the ground and shakes it, then inverts the bush, leaving the plant upside down on the ground to keep the peanuts out of the soil. This allows the peanuts to dry slowly to a little less than a third of their original moisture level over a period of three to four days. Traditionally, peanuts were pulled and inverted by hand.

After the peanuts have dried sufficiently, they are threshed, removing the peanut pods from the rest of the bush. It is particularly important that peanuts are dried properly and stored in dry conditions. If they are too high in moisture, or if storage conditions are poor, they may become infected by the mold fungus Aspergillus flavus. Many strains of this fungus release toxic and highly carcinogenic substances called aflatoxins.

Cultivation in China

Cultivars in the United States

There are many peanut cultivars grown around the world. The market classes grown in the United States are Spanish, Runner, Virginia, and Valencia. Peanuts produced in the United States are divided into three major areas: the southeastern United states region which includes Alabama, Georgia, and Florida; the southwestern United States region which includes New Mexico, Oklahoma, and Texas; and the third region in the general eastern United States which include Virginia, North Carolina, and South Carolina.

Certain cultivar groups are preferred for particular characteristics, such as differences in flavor, oil content, size, shape, and disease resistance. Most peanuts marketed in the shell are of the Virginia type, along with some Valencias selected for large size and the attractive appearance of the shell. Spanish peanuts are used mostly for peanut candy, salted nuts, and peanut butter.

Spanish group

The small Spanish types are grown in South Africa, and in the southwestern and southeastern US. Prior to 1940, 90% of the peanuts grown in Georgia, USA, were Spanish types, but the trend since then has been larger-seeded, higher-yielding, more disease-resistant cultivars. Spanish peanuts have a higher oil content than other types of peanuts. In the United States, the Spanish group is primarily grown in New Mexico, Oklahoma, and Texas.

Cultivars of the Spanish group include 'Dixie Spanish', 'Improved Spanish 2B', 'GFA Spanish', 'Argentine', 'Spantex', 'Spanette', 'Shaffers Spanish', 'Natal Common (Spanish)', "White Kernel Varieties', 'Starr', 'Comet', 'Florispan', 'Spanhoma', 'Spancross', 'OLin', 'Tamspan 90', 'AT 9899–14', 'Spanco', 'Wilco I', 'GG 2', 'GG 4', 'TMV 2', and 'Tamnut 06'.

Runner group

Since 1940, the southeastern US region has seen a shift to production of Runner group peanuts. This shift is due to good flavor, better roasting characteristics and higher yields when compared to Spanish types, leading to food manufacturers' preference for the use in peanut butter and salted nuts. Georgia's production is now almost 100% Runner type.

Cultivars of Runners include 'Southeastern Runner 56-15', 'Dixie Runner', 'Early Runner', 'Virginia Bunch 67', 'Bradford Runner', 'Egyptian Giant' (also known as 'Virginia Bunch' and 'Giant'), 'Rhodesian Spanish Bunch' (Valencia and Virginia Bunch), 'North Carolina Runner 56-15', 'Florunner', 'Virugard', 'Georgia Green', 'Tamrun 96', 'Flavor Runner 458', 'Tamrun OL01', 'Tamrun OL02' 'AT-120', 'Andru-93', 'Southern Runner', 'AT1-1', 'Georgia Brown', 'GK-7', and 'AT-108'.

Virginia group

The large-seeded Virginia group peanuts are grown in the US states of Virginia, North Carolina, Tennessee, Texas, New Mexico, Oklahoma, and parts of Georgia. They are increasing in popularity due to demand for large peanuts for processing, particularly for salting, confections, and roasting in the shells.

Virginia group peanuts are either bunch or running in growth habit. The bunch type is upright to spreading. It attains a height of 45 to 55 cm (18 to 22 in), and a spread of 70 to 80 cm (28 to 31 in), with 80-to-90 cm (31-to-35 in) rows that seldom cover the ground. The pods are borne within 5 to 10 cm of the base of the plant.

Cultivars of Virginia type peanuts include 'NC 7', 'NC 9', 'NC 10C', 'NC-V 11', 'VA 93B', 'NC 12C', 'VA-C 92R', 'Gregory', 'VA 98R', 'Perry', 'Wilson, 'Hull', 'AT VC-2' and' Shulamit'.

Valencia group

Valencia group peanuts are coarse, and they have heavy reddish stems and large foliage. In the United States, large commercial production is primarily in the South Plains of West Texas and eastern New Mexico near and south of Portales, New Mexico, but they are grown on a small scale elsewhere in the South as the best-flavored and preferred type for boiled peanuts. They are comparatively tall, having a height of 125 cm (49 in) and a spread of 75 cm (30 in). Peanut pods are borne on pegs arising from the main stem and the side branches. Most of the pods are clustered around the base of the plant, and only a few are found several inches away. Valencia types are three- to five-seeded and smooth, with no constriction of the shell between the seeds. Seeds are oval and tightly crowded into the pods. Typical seed weight is 0.4 to 0.5 g. This type is used heavily for sale roasted and salted in-shell peanuts and peanut butter. Varieties include 'Valencia A' and 'Valencia C'.

Tennessee Red and Tennessee White groups

These are alike, except for the color of the seed. Sometimes known also as Texas Red or White, the plants are similar to Valencia types, except the stems are green to greenish brown, and the pods are rough, irregular, and have a smaller proportion of kernels.

Production and trade

Peanuts are widely produced in tropical and subtropical regions of the world. China accounts for 37% of world production, India for 16%, Nigeria for 8%, and the United States for 6% (table). Major exporters are India, which accounts for 31% of world exports, the United States for 21%, Argentina for 10%, and the Netherlands for 8%. The European Union imports 32.5% of the world supply of shelled peanuts.

In the United States, Georgia is the leading peanut-producing state, followed by Texas and Alabama, respectively. About half of all peanuts produced in the US are grown within a 100 mi (160 km) radius of Dothan, Alabama. Dothan is home to the National Peanut Festival established in 1938 and held each fall to honor peanut growers and celebrate the harvest.

Food

Peanut Oil

Peanut oil is often used in cooking, because it has a mild flavor and a relatively high smoke point. Due to its high monounsaturated content, it is considered healthier than saturated oils, and is resistant to rancidity. The several types of peanut oil include: aromatic roasted peanut oil, refined peanut oil, extra virgin or cold-pressed peanut oil, and peanut extract. In the United States, refined peanut oil is exempt from allergen labeling laws.

Peanut Flour

Peanut flour is lower in fat than peanut butter, and has high protein content making it suitable as a flavor enhancer. Peanut flour is used as a gluten-free solution.

Boiled Peanuts

Boiled peanuts are a popular snack in the southern United States, as well as in India, China, and West Africa. In the US South, boiled peanuts are often prepared in briney water, and sold in streetside stands.

Dry-roasted peanuts

Dry peanuts can be roasted in the shell or shelled in a home oven if spread out one layer deep in a pan and baked at a temperature of 350 °F or 177 °C for 15 to 20 min (shelled) and 20 to 25 min (in shell).

Cuisine

Latin America

Peanuts are particularly common in Peruvian and Mexican cuisine, both of which marry indigenous and European ingredients. For instance, in Peru, a popular traditional dish is picante de cuy, a roasted guinea pig served in a sauce of ground peanuts (ingredients native to South America) with roasted onions and garlic (ingredients from European cuisine). Also, in the city of Arequipa in Peru, a well known dish called ocopa consists of a smooth sauce of roasted peanuts and hot peppers (both native to the region) with roasted onions, garlic, and oil, poured over meat or potatoes. Another example is a fricassee combining a similar mixture with sautéed seafood or boiled and shredded chicken. These dishes are generally known as ajíes, meaning "hot peppers", such as ají de pollo and ají de mariscos (seafood ajíes may omit peanuts). Likewise, during colonial times, the Spanish in Peru used peanuts to replace nuts unavailable in Peru, but used extensively in Spanish cuisine, such as almonds, pine nuts, and other nuts, typically ground or as paste and mixed with rice, meats, and vegetables for dishes such as rice pilaf.

Throughout the region, many candies and snacks are made using peanuts as a base.

Middle East

Crunchy coated peanuts, called kabukim in Hebrew, are a popular snack in Israel. Kabukim are commonly sold by weight at corner stores where fresh nuts and seeds are sold, though they are also available packaged. The coating typically consists of flour, salt, starch, lecithin, and sometimes sesame seeds. The origin of the name is obscure (it may be derived from kabuk which means nutshell or husk in Turkish). An additional variety of crunchy coated peanuts popular in Israel is "American peanuts". The coating of this variety is thinner, but harder to crack.

Bamba puffs are a popular snack in the Middle East. Their shape is similar to Cheez Doodles, but are made of corn and flavored with peanut butter.

Southeast Asia

Peanuts are also widely used in Southeast Asian cuisine, such as in Malaysia, Vietnam, and Indonesia, where they are typically made into a spicy sauce. Peanuts originally came to Indonesia from the Philippines, where the legume came from Mexico in times of Spanish colonization. One of the most famous Philippine dishes using peanuts is kare-kare, a mixture of meat and peanut butter.

Common Indonesian peanut-based dishes include gado-gado, pecel, karedok, and ketoprak, all vegetable salads mixed with peanut sauce, and the peanut-based sauce for satay.

South Asia

In the Indian subcontinent, peanuts are known as a light snack by themselves, usually roasted and salted (sometimes with the addition of chilli powder), and often sold roasted in pod, or boiled with salt. They are also made into little dessert or sweet snack pieces by processing with refined sugar and jaggery. Indian cuisine uses roasted, crushed peanuts to give a crunchy body to salads; they are added whole (without pods) to leafy vegetable stews for the same reason. Another use of peanut oil as cooking oil. Most Indians use mustard, sunflower, and peanut oil for cooking. In South india, groundnut 'chutney' is a popular combination, usually partaken with dosa and idli at breakfast.

West Africa

Peanuts grow well in southern Mali and adjacent regions of the Ivory Coast, Burkina Faso, Ghana, Nigeria, and Senegal; peanuts are similar in both agricultural and culinary qualities to the Bambara groundnut native to the region, and West Africans have adopted the crop as a staple. Peanut sauce, prepared with onions, garlic, peanut butter/paste, and vegetables such as carrots, cabbage, and cauliflower, can be vegetarian (the peanuts supplying ample protein) or prepared with meat, usually chicken. Peanuts are used in the Malian meat stew maafe. In Ghana, peanut butter is used for peanut butter soup nkate nkwan. Crushed peanuts may also be used for peanut candies nkate cake and kuli-kuli, as well as other local foods such as oto. Peanut butter is also an ingredient in Nigeria's "African salad".

Peanut powder is an important ingredient in the spicy coating for kebabs in Nigeria and Ghana.

East Africa

Peanuts are a common ingredient of several types of relishes (dishes which accompany nshima) eaten by the tribes in Malawi and in the eastern part of Zambia, and these dishes are now common throughout both countries. Thick peanut butter sauces are also made in Uganda to go with rice and other starchy foods. Groundnut stew, locally called ebinyebwa in Uganda, is made by boiling ground peanut flour with other ingredients, such as cabbage, mushrooms, dried fish, meat or other vegetables. Across East Africa, roasted peanuts, often in cones of newspaper, are a popular snack sold in the street.

North America

In Canada and the US, peanuts are used in candies, cakes, cookies, and other sweets. They are also enjoyed roasted and salted. Peanut butter is one of the most popular peanut-based foods in the US, and recipes for peanut soup are present in the United States south, Virginia in particular. In some southern portions of the US, peanuts are boiled for several hours until soft and moist. Peanuts are also deep-fried, shell and all.

Malnutrition

Peanuts are used to help fight malnutrition. Plumpy Nut, MANA Nutrition, and Medika Mamba are high-protein, high-energy, and high-nutrient peanut-based pastes developed to be used as a therapeutic food to aid in famine relief. The World Health Organization, UNICEF, Project Peanut Butter, and Doctors Without Borders have used these products to help save malnourished children in developing countries. Peanuts can be used like other legumes and grains to make a lactose-free, milk-like beverage, peanut milk, which is promoted in Africa as a way to reduce malnutrition among children.

Animal feed

Peanut plant tops are used for hay.

The protein cake (oilcake meal) residue from oil processing is used as an animal feed and as a soil fertilizer. Groundnut cake is a livestock feed, mostly used by cattle as protein supplements. It is one of the most important and valuable feed for all types of livestocks and one of the most active ingredient for poultry rations. Poor storage of the cake may sometimes results in its contamination by aflatoxin, a naturally occurring mycotoxin that is produced by Aspergillus flavus and Aspergillus parasiticus. The major constituents of the cake are essential amino acids such as lysine and glutamine. Other components are crude fiber, crude protein, and fat.

Industrial use

Peanuts have a variety of industrial end uses. Paint, varnish, lubricating oil, leather dressings, furniture polish, insecticides, and nitroglycerin are made from peanut oil. Soap is made from saponified oil, and many cosmetics contain peanut oil and its derivatives. The protein portion is used in the manufacture of some textile fibers. Peanut shells are used in the manufacture of plastic, wallboard, abrasives, fuel, cellulose (used in rayon and paper), and mucilage (glue).

Nutritional value

Peanuts are rich in essential nutrients (right table, USDA nutrient data). In a 100 g serving, peanuts provide 570 calories and are an excellent source (defined as more than 20% of the Daily Value, DV) of several B vitamins, vitamin E, several dietary minerals, such as manganese (95% DV), magnesium (52% DV) and phosphorus (48% DV), and dietary fiber (right table). They also contain about 25 g protein per 100 g serving, a higher proportion than in many tree nuts.

Some studies show that regular consumption of peanuts is associated with a lower risk of mortality specifically from certain diseases. However, the study designs do not allow cause and effect to be inferred. According to the US Food and Drug Administration, "Scientific evidence suggests but does not prove that eating 1.5 ounces per day of most nuts (such as peanuts) as part of a diet low in saturated fat and cholesterol may reduce the risk of heart disease."

Phytochemicals

Peanuts contain polyphenols, polyunsaturated and monounsaturated fats, phytosterols and dietary fiber in amounts similar to several tree nuts.

Peanut skins contain resveratrol which is under preliminary research for its potential effects in humans.

Oil composition

A common cooking and salad oil, peanut oil is 46% monounsaturated fats (primarily oleic acid), 32% polyunsaturated fats (primarily linoleic acid), and 17% saturated fats (primarily palmitic acid). Extractable from whole peanuts using a simple water and centrifugation method, the oil is being considered by NASA's Advanced Life Support program for future long-duration human space missions.

Groundnut

Vigna subterranea (also known by its common names Bambara groundnut, Bambara-bean, Congo goober, earth pea, ground-bean, or hog-peanut) is a member of the family Fabaceae. The plant originated in West Africa. Vigna subterranea ripens its pods underground, much like the peanut (also called a groundnut). They can be eaten fresh or boiled after drying.

Names in Other Languages

The name for the Bambara groundnut in the Bambara language of Mali (Bamanankan) is tiganing[l[n or tiganinkurun, literally "little hard peanut" (peanut is tiga). Among other West African languages it is known by other names; in Hausa it is variously referred to as Gurjiya or Kwaruru. In Goemai, Kwam; and in Kanuri, Ngangala; in Ibo, Okpa and in Ga, Akwei. In the Upper East Region of Ghana, it is referred to as "Suma".

Bambara groundnuts are also known as jugo beans or in Swahili, njugumawe. In the Republic of Zambia, bambara groundnuts are known as either Lituhu (siLozi), ntoyo (ciBemba),katoyo (kiKaonde), or mbwiila (chiTonga). In the Angolan plateau the plant is known as vi½lõ in the local Umbundu language, while in Angolan Portuguese it is mostly called feijão-gengibre (meaning ginger-bean). In Shangaan they are known as tindluwa. In Malagasy, the language of Madagascar, they are known as voanjobory, which translates to "round peanut." Interestingly enough, in the Ghanaian Ewe language, they are known as azi nogui, which also translates to "round peanut." Also, in the Dagbani language of the Dagombas in Northern Ghana, it is called 'simkpla', which also translates literally as 'round peanuts'. In Indonesia it is called "kacang bogor", which literally means "Bogor peanut", because this plant is widely cultured in Bogor, West Java. It is called "kacang poi" in Malay language, not to be confused with "kacang pool" which serves as one of popular dishes in the southern region of Malaysia. In the Shona language of Zimbabwe it is known as "Nyimo", and "indlubu" in the ndebele language. In Silozi, they are called 'Lituu.' In Oshiwambo language of Namibia, it is called "ofukwa" for sigular or "eefukwa" for plural.

Agronomic Aspects

Origin and Regions of Cultivation

The origin of the Bambara groundnut is West Africa and the region of cultivation is Sub-Saharan Africa's warm tropics.

Importance in the World Food System

Bambara groundnut represents the third most important grain legume in semi-arid Africa. "It is resistant to high temperature and is suitable for marginal soils where other leguminous crops cannot be grown". In addition, it makes very little demand on the soil and has a high nutritive value with 65% carbohydrate and 18% protein content. For these reasons it is not prone to the risk of total harvest failure even in low and uncertain rainfall regions. "Due to its high protein value it is a very important crop for people in Africa Despite its nutritional value, it is still considered as one of the prioritized neglected and underutilized species in Benin.

Culinary Use

The seeds are used for food and beverage because of its high protein content and for digestive system applications. The entire plant is known for soil improvement because of nitrogen fixation. In West Africa, the nuts are eaten as a snack, roasted and salted, or as a meal, boiled similar to other beans.

In South Eastern Nigeria, particularly in Enugu, the dried bambara beans are ground into a fine powder, then mixed with palm oil, water and pumpkin leaves and then poured into banana leaf wraps or 1 litre cellophane bags before being boiled into a kind of cakey pudding to make Enugu's most popular delicacy called "okpa" okpa is the number 1 breakfast item in Enugu, Nsukka and Ngwo Nigeria.

during the rainy season in many parts of central Nigeria like Jos, the fresh bambara beans are cooked with their shells still on them. the beans are then eaten as a snack just like boiled groundnuts.

Soil Requirements

Optimal soils for Bambara groundnut production are sandy soils to prevent waterlogging. Optimal soil depth is between 50 and 100 cm,with a light soil texture. soil fertility should be low and soil pH is best suited between 5 and 6.5 and should not be lower than 4.3or higher than 7.

Climate Requirements

The production is best suited between a latitude of 20° - 30°, i.e. the tropical wet and dry (Aw) and the subtropical dry summer (Cs) climate zones. Optimal temperature is between 19 °C and 30 °C. Temperatures below 16 °C and above 38 °C are not suited for the production of bambara groundnut. The bambara groundnut is very drought resistant. The minimal annual rainfall requirement is about 300 mm and optimal annual rainfall is between 750 mm and 1400 mm and should not exceed 3000 mm.

Cropping System

The cropping system is semi-permanent and the Bambara groundnut can be cultivated as single crop or as intercrop. Best suited intercrops are sorghum, millet,maize, groundnut, yams and cassava.

Bambara groundnut is mainly cultivated as intercrop, however the planting density varies between 6 and 29 plants per square meter. For woodland savannas of Côte d'Ivoire the highest yield is attainable with a plant density of 25 plants per square meter.

Fertilization

Since Vigna subterranea is usually intercropped, no fertilizer is applied. A yield of 1000 kg seed and 925 kg leaves remove 55.7 kg N, 26.2 kg K, 25.1 kg C, 7.8 kg Pand 6.6 kg Mg. Since Bambara groundnut is a legume, phosphor is the most important nutrient. An application rate of 60 kg/ha of P2O5 is recommended for bambara groundnut in Yola, Adamawa State Nigeria.

Sesame

Sesame is a flowering plant in the genus Sesamum, also called benne. Numerous wild relatives occur in Africa and a smaller number in India. It is widely naturalized in tropical regions around the world and is cultivated for its edible seeds, which grow in pods or "buns". The world harvested 4.2 million metric tonnes of sesame seeds in 2013, with India and China as the largest producers.Sesame seed is one of the oldest oilseed crops known, domesticated well over 3000 years ago. Sesame has many species, most being wild and native to sub-Saharan Africa. Sesame indicum, the cultivated type, originated in India and is tolerant to drought-like conditions, growing where other crops fail.Sesame has one of the highest oil contents of any seed. With a rich, nutty flavor, it is a common ingredient in cuisines across the world. Like other nuts and foods, it can trigger allergic reactions in some people.Sometimes sold with its seed coat removed (decorticated), this variety is often present on top of baked goods in many countries.Origins

Sesame seed is considered to be the oldest oilseed crop known to humanity. The genus has many species, and most are wild. Most wild species of the genus Sesamum are native to sub-Saharan Africa. Sesame indicum, the cultivated type, originated in India.

Charred remains of sesame recovered from archeological excavations have been dated to 3500-3050 BC. Fuller claims trading of sesame between Mesopotamia and the Indian sub-continent occurred by 2000 BC. Some reports claim sesame was cultivated in Egypt during the Ptolemiac period, while others suggest the New Kingdom.

Records from Babylon and Assyria, dating about 4000 years ago, mention sesame. Egyptians called it sesemt, and it is included in the list of medicinal drugs in the scrolls of the Ebers Papyrus dated to be over 3600 years old. Archeological reports from Turkey indicate that sesame was grown and pressed to extract oil at least 2750 years ago in the empire of Urartu.

The historic origin of sesame was favored by its ability to grow in areas that do not support the growth of other crops. It is also a robust crop that needs little farming support—it grows in drought conditions, in high heat, with residual moisture in soil after monsoons are gone or even when rains fail or when rains are excessive. It was a crop that could be grown by subsistence farmers at the edge of deserts, where no other crops grow. Sesame has been called a survivor crop.

Cultivation

Sesame varieties have adapted to many soil types. The high-yielding crops thrive best on well-drained, fertile soils of medium texture and neutral pH. However, these have low tolerance for soils with high salt and water-logged conditions. Commercial sesame crops require 90 to 120 frost free days. Warm conditions above 23 °C (73 °F) favor growth and yields. While sesame crops can grow in poor soils, the best yields come from properly fertilized farms.Initiation of flowering is sensitive to photoperiod and to sesame variety. The photoperiod also impacts the oil content in sesame seed; increased photoperiod increases oil content. The oil content of the seed is inversely proportional to its protein content.

Sesame is drought-tolerant, in part due to its extensive root system. However, it requires adequate moisture for germination and early growth. While the crop survives drought, as well as presence of excess water, the yields are significantly lower in either conditions. Moisture levels before planting and flowering impact yield most.

Most commercial cultivars of sesame are intolerant of water-logging. Rainfall late in the season prolongs growth and increases loss to dehiscence, when the seedpod shatters, scattering the seed. Wind can also cause shattering at harvest.

Processing

Sesame seeds are protected by a capsule which only bursts when the seeds are completely ripe. This is called dehiscence. The dehiscence time tends to vary, so farmers cut plants by hand and place them together in an upright position to continue ripening until all the capsules have opened. The discovery of an indehiscent mutant (analogous to nonshattering domestic grains) by Langham in 1943 began the work towards development of a high-yielding, dehiscence-resistant variety. Although researchers have made significant progress in sesame breeding, harvest losses due to dehiscence continue to limit domestic US production.

Since sesame is a small, flat seed, it is difficult to dry it after harvest because the small seed makes movement of air around the seed difficult. Therefore, the seeds need to be harvested as dry as possible and stored at 6% moisture or less. If the seed is too moist, it can quickly heat up and become rancid.

After harvesting, the seeds are usually cleaned and hulled. In some countries, once the seeds have been hulled, they are passed through an electronic colour-sorting machine that rejects any discolored seeds to ensure perfect colour. This is done because sesame seeds with consistent appearance are perceived to be of better quality by consumers, and sell for a higher price.

Immature or off-sized seeds are removed and used for sesame oil production.

Production and Trade

In 2013, world production of sesame seeds was 4.2 million tonnes, led by India and mainland China.

The most productive sesame seed farms in the world in 2013 were in Greece, reporting the highest nationwide yield of 0.69 tonnes per hectare. A large yield gap and farm loss differences exist between major sesame seed producers, in part because of knowledge gap, poor crop management practices, and use of technologies.

The white and other lighter-coloured sesame seeds are common in Europe, the Americas, West Asia, and the Indian subcontinent. The black and darker-coloured sesame seeds are mostly produced in China and southeast Asia. Africa produces a variety of sesame seeds.

Trade

The world traded over a billion dollars worth of sesame seeds in 2010. The trade volume has been increasing rapidly in the last two decades.

Japan is the world's largest sesame importer. Sesame oil, particularly from roasted seed, is an important component of Japanese cooking and traditionally the principal use of the seed. China is the second-largest importer of sesame, mostly oil-grade. China exports lower-priced food-grade sesame seeds, particularly to southeast Asia. Other major importers are the United States, Canada, the Netherlands, Turkey, and France.

Sesame seed is a high-value cash crop. Prices have ranged between US$800 and 1700 per metric ton between 2008 and 2010.

Sesame exports sell across a wide price range. Quality perception, particularly how the seed looks, is a major pricing factor. Most importers who supply ingredient distributors and oil processors only want to purchase scientifically treated, properly cleaned, washed, dried, colour-sorted, size-graded, and impurity-free seeds with a guaranteed minimum oil content (not less than 40%) packed according to international standards. Seeds that do not meet these quality standards are considered unfit for export and are consumed locally. In 2008, by volume, premium prices, and quality, the largest exporter was India, followed by Ethiopia and Myanmar.

Nutritional information

For a 100-gram serving, dried whole sesame seeds are rich in calories (573 kcal) and are composed of 5% water, 23% carbohydrates, 12% dietary fiber, 50% fat and 18% protein.

The flour that remains after oil extraction from sesame seeds is 35-50% protein and contains carbohydrates. This flour, also called sesame meal, is a high-protein feed for poultry and livestock.

Chemical composition

Sesame seeds contain the lignans, sesamolin, sesamin, pinoresinol and lariciresinol.

Cuisine

Sesame seed is a common ingredient in various cuisines. It is used whole in cooking for its rich, nutty flavour. Sesame seeds are sometimes added to breads, including bagels and the tops of hamburger buns. Sesame seeds may be baked into crackers, often in the form of sticks. In Sicily and France, the seeds are eaten on bread (ficelle sésame, sesame thread). In Greece, the seeds are also used in cakes.

Fast-food restaurants use buns with tops sprinkled with sesame seeds. About 75% of Mexico's sesame crop is purchased by McDonald'sfor use in their sesame seed buns worldwide.

In Asia, sesame seeds are sprinkled onto some sushi-style foods. In Japan, whole seeds are found in many salads and baked snacks, and tan and black sesame seed varieties are roasted and used to make the flavouring gomashio. East Asian cuisines, like Chinese cuisine, use sesame seeds and oil in some dishes, such as dim sum, sesame seed balls; Cantonese: jin deui), and the Vietnamesebánh rán. Sesame flavour (through oil and roasted or raw seeds) is also very popular in Korean cuisine, used to marinate meat and vegetables. Chefs in tempura restaurants blend sesame and cottonseed oil for deep-frying.

Sesame, or simsim as it is known in East Africa, is used in African cuisine. In Togo, the seeds are a main soup ingredient and in the Democratic Republic of the Congo and in the north of Angola, wangila is a delicious dish of ground sesame, often served with smoked fish or lobster.

Sesame seeds and oil are used extensively in India. In most parts of the country, sesame seeds mixed with heated jaggery, sugar, or palm sugar is made into balls and bars similar to peanut brittle or nut clusters and eaten as snacks. In Manipur, black sesame is used in the preparation of thoiding and in singju (a kind of salad). Thoiding is prepared with ginger and chili and vegetables are used in the spicy singu dish. In Assam, black sesame seeds are used to make til pitha and tilor laru (sesame seed balls) during bihu. In Punjab and Tamil Nadu, a sweet ball called pinni in Urdu and ell urundai in Tamil, ellunda in Malayalam, yellunde/Chigali, (sesame ball, usually in jaggery), is made of its seeds mixed with sugar. It is eaten in various forms during the festival of Makar Sankranti.

Also in Tamil Nadu, sesame oil used extensively in their cuisine, milagai podi, a ground powder made of sesame and dry chili is used to enhance flavor, and is consumed along with other traditional foods such as idli. In Tamil Nadu and Andhra Pradesh, sesame oil is used as a preservative, as well as to temper the heat of their spicy foods, pickles, and condiments. Sesame seed cookies and wafers, both sweet and savory, are popular in places such as Charleston, South Carolina. Sesame seeds, also called benne, are believed to have been brought into 17th-century colonial America by West African slaves. Since then, they have become part of various American cuisines.

In Caribbean cuisine, sugar and white sesame seeds are combined into a bar resembling peanut brittle and sold in stores and street corners.

Sesame is a popular and essential ingredient in many Middle Eastern cuisines. Sesame seeds are made into a paste called tahini (used in various ways, including hummus bi tahini) and the Middle Eastern confection halvah. Ground and processed, the seed is also used in sweet confections.

In South Asia, Middle East, and East Asian cuisines, popular confectionery are made from sesame mixed with honey or syrup and roasted into a sesame candy. In Japanese cuisine goma-dofu is made from sesame paste and starch.

Mexican cuisine refers to sesame seeds as ajonjolí. It is mainly used as a sauce additive, such as mole or adobo. It is often also used to sprinkle over artisan breads and baked in traditional form to coat the smooth dough, especially on whole-wheat flatbreads or artisan nutrition bars, such as alegrías.

In Sicilian cuisine, what are commonly called "Italian sesame seed cookies" are known as giuggiuleni . A giuggiulena usually refers to a cookie, while a giurgiulena usually refers to a nougat-like candy, often made as a Christmas food. Both are alternative spellings for "sesame seed" in the Sicilian language.

Sesame oil is sometimes used as a cooking oil in different parts of the world, though different forms have different characteristics for high-temperature frying. The "toasted" form of the oil (as distinguished from the "cold-pressed" form) has a distinctive pleasant aroma and taste, and is used as table condiment in some regions, especially in East Asia. Toasted sesame oil is also added to flavor soups and other hot dishes, usually just before serving, to avoid dissipating the volatile scents too rapidly.

Although sesame leaves are edible as a leaf vegetable, recipes for Korean cuisine calling for "sesame leaves" are often a mistranslation, and really mean perilla.

Allergy

Sesame seeds and sesame oil are a serious allergen to some people, including infants. In Australia, the occurrence of allergy to sesame seed was estimated to be 0.42% among all children, while in the United Kingdom, the allergic reaction was found to affect 0.04% of adults. The occurrence of allergy to sesame in patients with some form of food allergy was found to be much higher than in the general population, ranging from 0.5% in Switzerland to 8.5% in Australia. In other words, allergy to sesame affects a small percentage of overall human population, but sesame allergy is high in people who already show symptoms of allergy to other foods.

The symptoms of sesame seed allergy can be classified into:

- **Systemic reactions:** Primarily presenting anaphylaxis characterized by symptoms including hives (urticaria), lip and eyelid swelling (angioedema) sneezing, nasal itching, congestion, rhinorrhea, wheezing, cough, tightness of throat, hoarse voice, difficulty in breathing, abdominal pain, unconsciousness, shock with drop of blood pressure. In the systemic reactions can also be included severe reactions like dizziness, drowsiness, chills and collapse as has been reported in patients after ingestion of a falafel burger.
- **Other symptoms**: Facial or generalized redness ("flushing"), hives (urticaria) on smaller or larger parts of the body, swelling of the eyelids, lips or other parts of the face, itching of the eyes or of the skin in general, hay fever symptoms in the eyes and eczema. Respiratory symptoms observed include hay fever, asthma, cough, wheeze, or difficulty in breathing. Gastrointestinal symptoms: Itching in the mouth and/or tongue soon after chewing and ingesting (Oral allergy syndrome) and abdominal pain.

Amounts as low as 100 mg of sesame seeds or flour and 3 ml of oil can trigger allergic reactions in severe cases of sesame allergic individuals. Most patients, however, show allergic reactions after consuming 2–10 g of sesame seeds or flour. The onset of the symptoms may occur within a few minutes up to 90 minutes after ingestion of a sesame seed product. Most patients had other allergic diseases such as asthma, hay fever, and eczema, and most patients also had a relative with an allergic disease. More than two-thirds of the patients with sesame allergy also had allergic reactions to other foods.

Prevalence of sesame allergy varies per country. While it is one of the three most common allergens in Israel, sesame allergy prevalence is considered small relative to other allergens in the United States. Some experts consider sesame allergies to have "increased more than any other type of food allergy over the past 10 to 20 years" in the United States. Such increasing prevalence led Canada to issue regulations that require food labels to note the presence of sesame.

In addition to products derived from sesame such as tahini and sesame oil, persons with sesame allergies are warned to stay away from a broad assortment of processed foods, including baked goods, tempeh, and generic "vegetable oil". In addition to possible food sources, individuals allergic to sesame have been warned that a variety of nonfood sources may also trigger a reaction to sesame, including adhesive bandages, cosmetics, hair-care products, perfumes, soaps and sunscreens, drugs, some fungicides and insecticides, lubricants, ointments and topical oils, and pet food. At least one study found that "standard skin and blood testing for food allergies [does not] predict whether a child has true sesame allergy." In which case, a food challenge under the direction of a physician may be required to properly diagnose a sesame allergy.

A cross-reactivity apparently exists between sesame allergens and peanut, rye, kiwifruit, poppy seed, and various tree nuts (such as hazelnut, black walnut, cashew, macadamia, and pistachio).

Sunflower

Helianthus or sunflowers is a genus of plants comprising about 70 species in the family Asteraceae. The genus is one of many in the Asteraceae that are known as sunflowers. Except for three species in South America, all Helianthus species are native to North America. The common name, "sunflower", typically refers to the popular annual species Helianthus annuus, the common sunflower, whose round flower heads in combination with the ligules look like the sun. This and other species, notably Jerusalem artichoke (H. tuberosus), are cultivated in temperate regions as food crops and ornamental plants. The largest sunflower field is located in Tuscany, Italy.

The domesticated sunflower, H. annuus, is the most familiar species. Perennial sunflower species are not as popular for gardens due to their tendency to spread rapidly and become invasive. Whorled sunflowers, H. verticillatus, were listed as an endangered species in 2014 when the U.S. Fish and Wildlife Service issued a final rule protecting it under the Endangered Species Act. The primary threats are industrial forestry and pine plantations in Alabama, Georgia, and Tennessee. They grow to 1.8 m (6 ft) and are primarily found in woodlands, adjacent to creeks and moist, prairie-like areas.

Description

Sunflowers are usually tall annual or perennial plants that grow to a height of 300 centimetres (120 in) or more. They bear one or more wide, terminal capitula (flower heads), with bright yellow ray florets at the outside and yellow or maroon (also known as a brown/red) disc florets inside. Several ornamental cultivars of Helianthus annuus have red-colored ray florets; all of them stem from a single original mutant. During growth, sunflowers tilt during the day to face the sun, but stop once they begin blooming. This tracking of the sun in young sunflower heads is called heliotropism. By the time they are mature, sunflowers generally face east. The rough and hairy stem is branched in the upper part in wild plants but is usually unbranched in domesticated cultivars. The petiolate leaves are dentate and often sticky. The lower leaves are opposite, ovate or often heart-shaped.

They are distinguished technically by the fact that the ray florets (when present) are sterile, and by the presence on the disk flowers of a pappus that is of two awn-like scales that are caducous (that is, easily detached and falling at maturity). Some species also have additional shorter scales in the pappus, and there is one species that lacks a pappus entirely. Another technical feature that distinguishes the genus more reliably, but requires a microscope to see, is the presence of a prominent, multicellular appendage at the apex of the style. Sunflowers are especially well known for their symmetry based on Fibonacci numbers and the Golden angle.

There is quite a bit of variability among the perennial species that make up the bulk of the species in the genus. Some have most or all of the large leaves in a rosette at the base of the plant and produce a flowering stem that has leaves that are reduced in size. Most of the perennials have disk flowers that are entirely yellow, but a few have disk flowers with reddish lobes. One species, H. radula, lacks ray flowers altogether. Helianthus species are used as food plants by the larvae of many lepidopterans.

Safflower

Castor Oil

Castor oil is a vegetable oil obtained by pressing the seeds of the castor oil plant (Ricinus communis). The common name "castor oil", from which the plant gets its name, probably comes from its use as a replacement for castoreum, a perfume base made from the dried perineal glands of the beaver (castor in Latin).

Castor oil is a colorless to very pale yellow liquid with a distinct taste and odor once first ingested. Its boiling point is 313 °C (595 °F) and its density is 961 kg/m. It is a triglyceride in which approximately 90 percent of fatty acid chains are ricinoleates. Oleate and linoleates are the other significant components.

Castor oil and its derivatives are used in the manufacturing of soaps, lubricants, hydraulic and brake fluids, paints, dyes, coatings, inks, cold resistant plastics, waxes and polishes, nylon, pharmaceuticals and perfumes.

Composition

Castor oil is well known as a source of ricinoleic acid, a monounsaturated, 18-carbon fatty acid. Among fatty acids, ricinoleic acid is unusual in that it has a hydroxyl functional group on the 12th carbon. This functional group causes ricinoleic acid (and castor oil) to be more polar than most fats. The chemical reactivity of the alcohol group also allows chemical derivatization that is not possible with most other seed oils. Because of its ricinoleic acid content, castor oil is a valuable chemical in feedstocks, commanding a higher price than other seed oils. As an example, in July 2007, Indian castor oil sold for about US$0.90 per kilogram (US$0.41 per pound) whereas U.S. soybean, sunflower and canola oils sold for about US$0.30 per kilogram (US$0.14 per pound).

Average composition of castor seed oil / fatty acid chains

Acid name	Average Percentage Range
Ricinoleic acid	85 - 95%
Oleic acid	2 - 6%
Linoleic acid	1 - 5%
a-Linolenic acid	0.5 - 1%
Stearic acid	0.5 - 1%
Palmitic acid	0.5 - 1%
Dihydroxystearic acid	0.3 - 0.5%
Others	0.2 - 0.5%

Uses

Annually 270,000–360,000 tonnes (600–800 million pounds) of castor oil are produced for a variety of uses.

Food and Preservative

In the food industry, castor oil (food grade) is used in food additives, flavorings, candy (e.g., polyglycerol polyricinoleate or PGPR in chocolate), as a mold inhibitor, and in packaging. Polyoxyethylated castor oil (e.g., Kolliphor EL) is also used in the food industries.

In India, Pakistan and Nepal food grains are preserved by the application of castor oil. It stops rice, wheat, and pulses from rotting. For example, the legume pigeon pea is commonly available coated in oil for extended storage.

Medicine

The United States Food and Drug Administration (FDA) has categorized castor oil as "generally recognized as safe and effective" (GRASE) for over-the-counter use as a laxative with its major site of action the small intestine where it is digested into ricinoleic acid. Despite castor oil being widely used to induce labor in pregnant women, to date there is not enough research to show whether it is effective to ripen the cervix or induce labor. Therapeutically, modern drugs are rarely given in a pure chemical state, so most active ingredients are combined with excipients or additives. Castor oil, or a castor oil derivative such as Kolliphor EL (polyethoxylated castor oil, a nonionic surfactant), is added to many modern drugs, including:

- Miconazole, an antifungal agent;
- Paclitaxel, a mitotic inhibitor used in cancer chemotherapy;
- Sandimmune (cyclosporine injection, USP), an immunosuppressant drug widely used in connection with organ transplant to reduce the activity of the patient's immune system;
- Nelfinavir mesylate, an HIV protease inhibitor;
- Tacrolimus, an immunosuppressive drug (contains HCO-60, polyoxyl 60 hydrogenated castor oil);
- Xenaderm ointment, a topical treatment for skin ulcers, is a combination of Balsam of Peru, castor oil, and trypsin;
- Aci-Jel (composed of ricinoleic acid from castor oil, with acetic acid and oxyquinoline) is used to maintain the acidity of the vagina.
- Optive Plus (carboxymethylcellulose, castor oil) and Refresh Ultra (glycerine, castor oil), are artificial tears to treat dry eye.

Castor oil is also one of the components of Vishnevsky liniment.

Alternative Medicinal Use

In naturopathy castor oil has been promoted as a treatment for a variety of human health conditions, including cysts. The claim has been made that applying it to the skin can help cure cancer. However, according to the American Cancer Society, "available scientific evidence does not support claims that castor oil on the skin cures cancer or any other disease."

Coatings

Castor oil is used as bio-based polyol in the polyurethane industry. The average functionality (number of hydroxyl groups per triglyceride molecule) of castor oil is 2.7, so it is widely used as rigid polyol and coating.

Castor oil is not a drying oil, meaning that it has a low reactivity toward air compared to other oils such as linseed oil and tung oil. Dehydration of castor oil gives linoleic acids, which do have drying properties.

Precursor to Industrial Chemicals

Castor oil can be broken down into other chemical compounds that have numerous applications. Transesterification followed by steam cracking gives undecylenic acid, a precursor to specialized polymer nylon 11, and heptanal, a component in fragrances. Breakdown of castor oil in strong base gives 2-octanol, both a fragrance component and a specialized solvent, and the dicarboxylic acid sebacic acid. Hydrogenation of castor oil saturates the alkenes, giving a waxy lubricant.

- The production of lithium grease consumes a significant amount of castor oil. Hydrogenation and saponification of castor oil yields 12-hydroxystearic acid which is then reacted with lithium hydroxide or lithium carbonate to give high performance lubricant grease.
- Since it has a relatively high dielectric constant (4.7), highly refined and dried castor oil is sometimes used as a dielectric fluid within high performance high voltage capacitors.

Lubrication

Vegetable oils like castor oil are typically unattractive alternatives to petroleum-derived lubricants because of their poor oxidative stability. Castor oil has better low-temperature viscosity properties and high-temperature lubrication than most vegetable oils, making it useful as a lubricant in jet, diesel, and racing engines. The viscosity of castor oil at 10 °C is 2,420 centipoise. However, castor oil tends to form gums in a short time, and therefore its usefulness is limited to engines that are regularly rebuilt, such as racing engines. The lubricants company Castrol took its name from castor oil.

Castor oil has been suggested as a lubricant for bicycle pumps because it does not degrade natural rubber seals.

Early aviation and aeromodelling

Castor oil was the preferred lubricant for rotary engines, such as the Gnome engine after that engine's widespread adoption for aviation in Europe in 1909. It was used almost universally in rotary engined Allied aircraft in World War I. Germany had to make do with inferior ersatz oil (a German term for alternative, substitute, replacement) for its rotary engines, which resulted in poor reliability.

The methanol-fueled two-cycle glow plug engines used for aeromodelling, since their adoption by model airplane hobbyists in the 1940s, have used varying percentages of castor oil as a lubricant. It is highly resistant to degradation when the engine has its fuel-air mixture leaned for maximum engine speed. Gummy residues can still be a problem for aeromodelling powerplants lubricated with castor oil, however, usually requiring eventual replacement of ball bearings when the residue accumulates within the engine's bearing races. One British manufacturer of sleeve valved four-cycle model engines has stated the "varnish" created by using castor oil in small percentages can improve the pneumatic seal of the sleeve valve, improving such an engine's performance over time.

Turkey red oil

Turkey red oil, also called sulphonated (or sulfated) castor oil, is made by adding sulfuric acid to vegetable oils, most notably castor oil.It was the first synthetic detergent after ordinary soap. It is used in formulating lubricants, softeners, and dyeing assistants.

Biodiesel

Castor oil, like currently less expensive vegetable oils, can be used as feedstock in the production of biodiesel. The resulting fuel is superior for cold winters, because of its exceptionally low cloud and pour points.

Initiatives to grow more castor for energy production, in preference to other oil crops, are motivated by social considerations. Tropical subsistence farmers would gain a cash crop.

Punishment

Parents often punished children with a dose of castor oil. Physicians recommended against the practice because they did not want medicines associated with punishment.

A heavy dose of castor oil could be used as a humiliating punishment for adults, especially political dissenters. Colonial officials used it in the British Raj (India) to deal with recalcitrant servants.

The most famous use as punishment came in Fascist Italy under Benito Mussolini. It was a favorite tool used by the Blackshirts to humiliate their opponents. Political dissidents were force-fed large quantities of castor oil by Fascist squads. This technique was said to have been originated by Gabriele D'Annunzio. Victims of this treatment did sometimes die, as the dehydrating effects of the oil-induced diarrhea often complicated the recovery from the nightstick beating they also received along with the castor oil; however, even those victims who survived had to bear the humiliation of the laxative effects resulting from excessive consumption of the oil.

It is said that Mussolini's power was backed by "the bludgeon and castor oil". In lesser quantities, castor oil was also used as an instrument of intimidation, for example, to discourage civilians or soldiers who would call in sick either in the factory or in the military. It took decades after Mussolini's death before the myth of castor oil as a panacea for a wide range of diseases and medical conditions was totally demystified, as it was also widely administered to pregnant women, and elderly or mentally ill patients in hospitals in the false belief it had no negative side effects.

Today, the Italian terms manganello and olio di ricino, even used separately, still carry strong political connotations (especially the latter). These words are still used to satirize patronizing politicians, or the authors of disliked legislation. They should be used with caution in common conversation. The terms Usare l'olio di ricino, ("to use castor oil") and usare il manganello ("to use the bludgeon") mean "to coerce or abuse", and can be misunderstood in the absence of proper context.

As a means of punishment or torture, force-feeding castor oil still lives on in animated cartoons such as Tom and Jerry.

Safety

The castor seed contains ricin, a toxic protein. Heating during the oil extraction process denatures and deactivates the protein. However, harvesting castor beans may not be without risk. Allergenic compounds found on the plant surface can cause permanent nerve damage, making the harvest of castor beans a human health risk. India, Brazil, and China are the major crop producers, and the workers suffer harmful side effects from working with these plants. These health issues, in addition to concerns about the toxic byproduct (ricin) from castor oil production, have encouraged the quest for alternative sources for hydroxy fatty acids. Alternatively, some researchers are trying to genetically modify the castor plant to prevent the synthesis of ricin.

While castor oil is sometimes used to induce labor in full-term pregnancies (scientific evidence of its effectiveness is lacking), consuming castor oil to treat constipation is not considered safe in pregnancies that are not at full term yet, since it may cause contractions of the womb.

Mustard plant

Mustard plants are any of several plant species in the genera Brassica and Sinapis. Mustard seed is used as a spice. Grinding and mixing the seeds with water, vinegar, or other liquids, creates the yellow condiment known as prepared mustard. The seeds can also be pressed to make mustard oil, and the edible leaves can be eaten as mustard greens.

History

Although some varieties of mustard plants were well-established crops in Hellenistic and Roman times, Zohary and Hopf note: "There are almost no archeological records available for any of these crops." Wild forms of mustard and its relatives, the radish and turnip, can be found over west Asia and Europe, suggesting their domestication took place somewhere in that area. However, Zohary and Hopf conclude: "Suggestions as to the origins of these plants are necessarily based on linguistic considerations." Encyclopædia Britannica states that mustard was grown by the Indus Civilization of 2500-1700 BCE, According to the Saskatchewan Mustard Development Commission, "Some of the earliest known documentation of mustard's use dates back to Sumerian and Sanskrit texts from 3000 BC"

Varieties

Mild white mustard (Sinapis hirta) grows wild in North Africa, the Middle East, and Mediterranean Europe, and has spread farther by long cultivation; oriental mustard (Brassica juncea), originally from the foothills of the Himalaya, is grown commercially in India, Canada, the UK, Denmark, and the US; black mustard (Brassica nigra) is grown in Argentina, Chile, the US and some European countries. Canada and Nepal are the world's major producers of mustard seed, between them accounting for around 57% of world production in 2010. White mustard (Sinapis alba) is commonly used as a cover crop in Europe (between UK and Ukraine). A large number of varieties exist, e.g. in Germany, Netherlands, mainly differing in lateness of flowering and resistance against white beet-cyst nematode (Heterodera schachtii). Farmers prefer late flowering varieties, which do not produce seeds, they may become weeds in the subsequent year. Early vigour is important to cover the soil quickly and suppress weeds and protect the soil against erosion. In rotations with sugar beets, suppression of the white beet-cyst nematode is an important trait. Resistant white mustard varieties decline nematode populations by 70-90%.

Recent research has studied varieties of mustards with high oil contents for use in the production of biodiesel, a renewable liquid fuel similar to diesel fuel. The biodiesel made from mustard oil has good cold flow properties and cetane ratings. The leftover meal after pressing out the oil has also been found to be an effective pesticide.

10

HORTICULTURAL GENETIC RESOURCE MANAGEMENT

Sustainable, resilient horticultural systems and improved nutrition depend on the conservation and sustainable use of crop diversity. Promoting the sustainable production and consumption of horticultural diversity can improve the livelihoods and nutritional status of both urban and rural dwellers. Many horticultural crops are sidelined by mainstream agriculture today, in spite of their potential contribution to people's livelihoods, their nutritional value and potential role in climate change adaptation. In addition, these resources are being eroded through changes in agricultural practices and in dietary preferences, destruction or fragmentation of habitats, climate change and other factors.

There is, therefore, a pressing need to conserve horticultural diversity on-farm as well as securing their crop wild relatives and wild edible fruits and vegetables in situ and in complementary ex situ programs. Surveying, inventorying, and monitoring are the key initial steps for identifying, prioritizing and managing horticultural genetic resources. The development of novel, well-adapted and productive crops is important for pre-breeding programs and the selection of genetic material. These activities require promoting increased capacities across the different stakeholders who contribute to the conservation and sustainable use of horticultural genetic resources. There is also a need to integrate traditional management practices and ethnobotanical approaches while promoting the development of novel approaches and technologies.

Linking these elements together, streamlined strategies and policies are necessary for ensuring the conservation and sustainable use of these resources while promoting partnerships at national, regional and international levels is crucial for greater impact and uptake of best practices.

Global climate change is causing challenges for horticultural crop productivity with the constant rise in global temperatures and reduced amount of precipitation. The proper utilization of genetic resources can help develop resilient horticultural varieties to adopt to changing environmental conditions. The specifications entail a necessary agricultural adaptation to climate change.

It is crucial to develop strategies for effective conservation and judicious use of the genetic resources of horticultural crops to improve the nutritional security and food safety for human health. A large variation in species diversity contributes to the conservation of valuable resources within each species of fruits, vegetables, medicinal, aromatic, and ornamental plants. The importance of horticultural biodiversity is widely documented, emphasizing alleviating the adverse effects of climate changes on the biodiversity. The introgression of local germplasms and their conservation helps obtain new varieties with a more comprehensive genetic background to enable suitable production under different conditions and elicit appropriate responses to various types of stress. Therefore, identifying and evaluating genetic resources would be a vital prerequisite for horticultural crop production.

While these conventional approaches can be accomplished by the direct utilization of genetic resources in breeding programs, recent advancements in biotechnology have progressively used various tools for selecting potential parents from germplasm collections. Some of these tools are embedded in specialized techniques and biotechnological assets, such as next-generation sequencing (NGS), high-throughput genotyping platforms (SNP genotyping array and genotyping by sequencing), and genomics-based approaches such as genome-wide association studies (GWAS), marker-assisted selection (MAS), and genomic selection (GS). These methods also include relevant Omics approaches such as transcriptomics, proteomics and metabolomics in generating new cultivars with higher adaptabilities to environmental conditions.

Given the context of this research topic, we aim to cover a wide range of studies on germplasm evaluation and conservation, domestication of horticultural crops, population genomics, climatic change and adaptive evolution. We welcome original research papers, mini-reviews, reviews, and short communications on topics that include but are not limited to:

- Utilization of landraces and wild relatives of plants to establish adaptation to climate change
- Application of genetic resources to develop desirable cultivars or varieties
- Exploration and conservation of genetic diversity in horticultural crops
- Geographical and environmental variation in chemical constituents and antioxidant properties in medicinal plants
- Understanding physiological and biochemical mechanisms of abiotic stress tolerance in genetic resources of horticultural crops using traditional and modern phenotyping techniques

- Understanding molecular mechanisms underlying the different types of abiotic stress tolerance in genetic resources of horticultural crops through diverse facets of the multi-omics approach that involve genomics, transcriptomics, proteomics and metabolomics

Present status of PGR activities at IIHR Bangalore

The PGR Division at IIHR Bangalore since its inception in 1978 has played an important role in crop improvement and diversification of horticulture in India through acquisition and distribution of the germplasm of fruits, vegetables, ornamentals and Medicinal and Aromatic Plant species in liaison with NBPGR. From being a section primarily concerned with plant introduction, exchange and conservation of gene pool components such as seed and pollen, PGR division today plays a key role in the overall management of horticultural PGR, comprising activities such as germplasm exploration, collection, exchange, quarantine, characterization, evaluation, conservation and documentation. The division was initially funded through a UNDP program for HR development and initiation of research in the area of germplasm conservation.

The Division has built up infrastructural facilities in the form of a full fledged seed storage laboratory, pollen storage laboratory having facilities for Cryopreservation of horticultural PGR. In 1996, the division developed facilities for in vitro conservation through an AP Cess funded project for optimizing in vitro conservation protocols for grape germplasm. Further support was provided to the division from 1999-2003 through an NATP Plant Biodiversity project, during which period the cryogenic and in vitro conservation facilities were strengthened, besides receiving support for plant germplasm exploration of species of relevance to Horticulture, from the Western Ghat region. It was during this period, IIHR developed a FGB facility for over 100 species of horticultural importance, well recognized and appreciated by NBPGR, which also led to identification of IIHR as NAGS for tropical & subtropical fruits.

The Division was is also carrying out its mandatory baseline PGR activities related to introduction and exchange indigenous and exotic germplasm of horticultural crops in liaison with NBPGR.

Achievements so far

Plant germplasm is an important natural resource and plays a vital role in the sustainable development of horticulture. A wide genetic base is essential for the development of new genotypes capable of surviving under heterogeneous environment. Management of genetic resources draws a major attention in farming and current emphasis is placed especially in areas such as exploration and collection, introduction, evaluation, characterization, documentation, and conservation of valuable genetic diversity in different horticultural crops. Keeping in view the importance of plant genetic resources, the section of Plant Genetic Resources has been upgraded to the Division level in 1999 to give fillip to the research in the area of genetic of resources management. Since then, the division has been working to evolve strategies for sustainable management of genetic resources of various horticultural crops, comprising fruits, vegetables, ornamental and medicinal crops.

I. Exploration and Collection of Germplasm

Exploration and collection missions were undertaken covering various parts of Southern India to collect wild as well as cultivated but untapped genetic resources of various horticultural crops.

1. Exploration and collection mission to Maharashtra

The districts covered were Pune, Satara, Kolhapur and Rathnagiri. A total no of 50 accessions of cowpea, dolichos, French bean, amaranth, brinjal and cucurbits were collected. A wild relative of the Vigna unguiculata (Vigna marina) was collected from Shivaji University, Kolhapur (it is distributed in Nicobar island). Collection from Gowal village in Rathnagiri district is cultivating specialty vegetable in cowpea and brinjal and germplasm of the same were collected.

2. Exploration and collection mission to Dharwad region of Karnataka

Places of collection were Dharwad, Gadag, districts and particularly Hulikoti, Gokak, Ghataprabha, Kabbur and Hukkeri market. A total of 28 germplasm were collected including amaranth, brinjal, dolichos, cowpea, cucurbits, tomato, and okra. Many local varieties of brinjal like Malapur local, Kuduchi local, Betegere local were collected. A Visit was made to growers field in Banavasi, Sirsi, Uttara Kannada, and collected four different lines belonging to Kew and Queen cultivars. Two variants in Kew cultivar, one brought from Punalur and the other from Thrissur were also collected. A survey was conducted in and around Sirsi, Sagar, Siddapur and Ullur in North Kannada district for mango trees used for pickling. Some of the lines collected (local names) are Kadigai, Karishad, Chandrama, Kempigunde, Gundappe, Maitri, Gangappe, Magemavu, Kadakeri local, Chinithotta, Dombesara, Kelaginamane, Yellare, Siddapura, Jeerige midi, Kanchape, karpurappe and Saadappe.

3. Exploration and collection mission to Tamil Nadu and Pondicherry

Districts covered were Cuddalore, Villupuram and Pondicherry union territory. Many local varieties of brinjal were collected viz. Annamalai University local, Palur local. In some pockets of pondicherry district, local brinjal and local red onion were also collected.

A visit to Parassara forest area resulted in collection of two pineapple lines and at HRS, Pechiparai, collections of two lines of pineapple lines one cultivar of jack (PPI-1) and Myristica fragrans were made.

An exploratory mission to Tirunelvi was undertaken for survey of off season mango and medicinal plants covering Parassala, Kanyakumari, Tirunelveli, Tenkasi,

Tuthukudi, Vallanad and surrounding areas. At Prassala, a local mango variety namely Chengavarikkai was collected. From Hoticulture farm, Kanyakumari, scion sticks of local varieties such as Surankudi, Kalkaichi, Nadan, Mohandas, Vettaiyasurankudi, Chinnakkothu, Panchavarnam, Kalkandu and Panickanadan were collected. Four established varieties of aonla were also procured. One medicinal plant locally known as Vaadhamudakki was collected. Two sets of Tirunelveli Senna were collected from Senna Exporters of Tirunelveli. Exploration at Vallanad hills resulted in location of one line each in Aloe and Shatavari. At Ambasamudram, one line each of glory lily and Hemedesmus were also collected.

4. Exploration and collection mission to Kerala

An exploration mission was undertaken to Northern Kerala covering districts of Calicut, Malappuram of Palghat to collect indigenous ornamental germplasm. The species collected were Alpinia calcarata, Selaginella lepidophylla, Kaempferia rotunda, Cholorophytum laxatum Curcuma longa, Cucurma oligathis and Globba vivipara, Alpinia species, Curcuma aromatica, Curcuma cesia Costus speciosus ,Kaempferia species, Hedichium coronatrium, Curcuma aruginosa Cucurma ecalcarata, Alpinia calcarata and Alpinia zerumbet .

A mission was undertaken to Thiruvananthapuram and Kanyakumari to collect fruit and medicinal plants. Exploration to forests around Palaruvi, Tenmalai and Ariyankavu resulted in collection of Gardenia gummifera, Andrographis wightiana, Rauwolfia serpentine, Stychna calibryanum, Cycas circinalis, Alstonia veninata, Cinnamomum malabaricum, Elephantopus scabet, Piper longum, Mangifera indica and Celasterous paniculata. Procured Aloe vera (3 types), Bacopa monneri (4 accessions), Centella asiatica (5 accessions), Piper betel (3 accessions), Andrographis sp., Plumbago rosea, Asparagus racemosus and seeds of Gloriosa superba, Andrographis paniculata, Garcinia gummigatta and Mucuna grandis from TBGRI, Thiruvanathapuram. Lines of lemon grass, Asparagus racemosus, Jatropha glandulifera, Mucuna pruriens, Mucuna atropurpuria, Mucuna cochinensis and Stychna potatarum were also procured.

A total of 12 fruit species common to South Kerala namely, Antidesma ghaesmbilla (Black current), Salacia chinensis, Salacia beddomei, Flacourtia Montana, Syzygium zeylanicum, Elaecarpus serratus, Chrysophyllum cainito (Star apple), Flacourti (lovi-lovi), Averrhoa bilimbi (Bilimbi), Averrhoa carambola (sour type Carambola), Psidium guineense, Aporusa lindleyana and Psidium littorale, from forests were procured from TBGRI.

The exploration to Vaniampuzha forest area resulted in collection of Cyclea peltata, Asparagus racemosus, Gloriosa superba, Mucuna pruriens, and Piper longum. Suckers of Aloe abyssina were procured from the Herbal garden of KFRI. Collected plants and tubers of Aloe from breeding block of KAU and farmers field.

5. Exploration and collection mission to Andaman and Nicobar Islands

Fruits of Mangifera andamanica and M. camptosperma were collected. The other species collected were wild Jamun, wild artocarpus, Ciba pentandra (Silk cotton tree) and Oroxylum indicum, large oval shaped fruits of Aegle marmelos. Fruits of Mangifera griffithi and Garcinia sp were collected. Two plants each of Myristica andamanica and Cycas rhumphi were also collected. A naturally growing Mucuna plant was also located and dried fruits were collected. Betel vine plants were collected which looked like cultivated as well as wild. The leaves were more pungent to chew. Also several medicinal plants were collected.

6. Exploration and collection mission to Goa

An extensive survey of the North Goa covering Balki, Savai Verum, Borin, Khandapar, Marcela, Khandola and Bicholim was undertaken. Kokam trees were found to be fairly distributed all over the jungles of Goa with wide variations in vegetative and fruit characteristics. In all, 33 trees were located and studied in detail. Fruit samples were collected for further studies. Based on preliminary observations, two trees could be considered as elite trees.

7. Exploration and collection mission to parts of Karnataka

Exploration was carried out in the Devarayana Durga forest range of the Tumkur district of Karnataka and 18 medicinal plants were collected. Similarly, 12 medicinal plants were collected from Koodluthirtha and Agumbe area. Kokam trees were collected from Puttur in Dakshin Kannada district. A Survey was undertaken to Puttur, South Canara to collect variability in Kokum trees. Five different hermaphrodite trees were located and studied from private orchards in Alakku Majalu, Marial Puttur and Savanuru. Two elite trees with big sized very dark purple coloured fruits and high yield and one tree with early bearing habit were collected from a private orchrd at Sediyapu, which has plenty of variability for kokum trees.

8. Collection mission to Andhra Pradesh

A Survey was undertaken to Adilabad, AP and collected fruits and scion sticks of Gourani mango, which has good canopy and good fruit yield from Dongragaon of Gudihathnoor mandal. In Narayanpur village collected Gourani types with pinkish shoulder, golden yellow and very early maturing types. A similar survey was undertaken to Emboye village of Bethamcherla (AP), known for excellent quality Banganapalli and collected local varieties Narasu and Reddipasand, one of which is claimed to be better than Banganapalli in quality. Fruits and scion sticks of four such types including golden yellow colour with good quality fruits and another with purplish red colour were collected.

Genetic diversity analysis

French bean (Phaseolus vulgaris L.)

The molecular diversity of French bean 48 accessions collected from 7 different agro-climatic zones of India were evaluated using RAPD markers. The molecular dissimilarity values ranged from 4 to 45% suggesting a narrow genetic base within the Phaseolus vulgaris cultivars. Principal component analysis showed two major clusters similar to the dendrogram of cluster analysis.

Onion

The genetic diversity of onion genotypes and hybrids were assessed using DNA markers. Twenty four breeding lines of onion were selected based on colour of the bulb, disease resistance, bulb type (single or multiplier), type of pollination and hybrids. Randomly selected eleven and thirteen sets of breeding lines were studied separately to know the genetic similarity using DNA markers. For the eleven breeding lines, seven random primers generated 42 RAPD loci of which 35 RAPD loci were polymorphic (81.89%). Among the thirteen breeding lines, five random primers generated 37 RAPD loci of which 33 RAPD loci were polymorphic (86.66%). Out of the seven primers OPA 18 showed the highest genetic polymorphism of 100% indicating that the primer can be used for fingerprinting. In the other thirteen breeding lines, OPC 05 and OPC 06 produced 100% polymorphism. Each cluster in the dendrograms belonged to their respective groups. The information generated can be used to improve the agronomically important traits.

Ornamental Crops: DNA fingerprinting of released varieties was done for the following ornamental crops using ISSR/SSR markers to establish genetic distinctness of the varieties and to protect plant breeder's rights. The following varieties have been characterized using DNA markers.

Rose: DNA profiles of three Varieties viz., G S Randhawa, Kiran and Nishkant developed using SSR and ISSR markers

Crossandra : DNA profiles of Arka kanaka, Arka ambara, 2005-1 and 2005-2 generated using ISSR markers..

Carnation: DNA Fingerprinting of Variety IIHR P-1 (Arka Flame) was carried out using SSR markers

Tuberose: DNA profiles of Sringar, Suvasini, Prajwal, Vybhav, Arka Niranthara were generated using ISSR markers.

China Aster: DNA profiles of Poornima, Shashank, Kamini were generated using ISSR markers.

Gladiolus: Molecular characterization and genetic diversity analysis carried out for 30 Genotypes. Cluster analysis based on Ward's method showed two major clusters. First cluster consisted of IIHRG-4, IIHRG- 9, IIHRG- 5, Aarthi, Sanjeeveni, Tilak, Dr.Flemming, Wild Rose, Shobha, Shakti, IIHRG -6, IIHRG -8 and IIHRG – 12. The second cluster consists of Apsara, Basant bahar, Arka kesar, Poonam, Sapna, Sagar, Sindhur, Darshan, Kum Kum, Gladiolus callianthus, IIHRG- 11, Shirley, Psittacinus (local) , IIHRG- 10, Pink friendship, Candy man and Dhanvantri. Genotypes Dhanvantri and candyman can be considered as potent parents for further improvement programme. The polymorphic primers identified provide a core set that will be useful for fingerprinting gladiolus cultivars/hybrids/species. Gladiolus callianthus, a fragrant genotype can be utilised for breeding programs.

Medicinal Crops:

Mucuna::Genetic diversity analysis has been carried out in the accessions of Mucuna pruriens. RAPD profiles of the 22 accessions taken for study were generated with four random primers and its pair wise combinations. The primer generated 152 RAPD loci of which 113 were polymorphic. The level of polymorphism generated was 74% among the accessions. Among the individual primers used OPA 2 produced maximum polymorphism of 65% and in pair wise combination OPA1+ OPA 2 generated 94 % polymorphism. It was found that adopting the pair wise primer combination enhanced the level of polymorphism as compared to the individual primer. The dendrogram based on Ward's method grouped the 22 accessions into four clusters and one lone cluster. The lone cluster belonged to the genotype MP-5 that has a high L-DOPA content.

Identification of Zygotic seedlings using molecular Markers

Mango: Distinguishing zygotic seedling in polyembryonic (PE) varieties of mango using molecular markers is in progress in collaboration with the division of Fruit crops.

Minor fruit crops: Jack fruit (Artocarpus heterophyllus): Molecular characterization and genetic diversity analysis in under utilized fruit crop namely jack fruit (Artocarpus heterophyllus) is in progress for identifying trait specific germplasm to establish working collections.

Genetic stability/fidelity testing using molecular markers.

PCR protocols were optimised and identified ISSR/SSR primers for genetic fidelity testing of tissue culture raised plantlets of Anthurium, Carnation and Gerbera.

Genetic fidelity of in vitro raised plantlets of carnation vis-a-vis mother plant was assessed using 16 SSR and 10 ISSR. The percentage of polymorphic bands in SSR and ISSR analysis were respectively 7.69% and 5.88% for the 16 clones and mother plant. Microsatellite markers detected more polymorphism in the in vitro clones compared to ISSR markers revealing the ability of these markers to detect subtle variation at the molecular level.

Germplasm Evaluation

A field evaluation trial indicated that Bird of Paradise is a promising crop for southern dry zones of Karnataka. Identified Bird of Paradise Elite lines with specific attributes, such as plant height, spike length, high flower to leaf ratio, clump production, high spike yield and more florets/spike, which can be used as clonal selections or used for as potential parents for crop improvement programmes.

Generated mutant population in bird of paradise by gamma irradiation for increasing variability and isolating mutants with desirable traits. Initial evaluation indicated a wide range of variability with respect to plant height, number of leaves/plant and clump production and spike yield.

Ii. Germplasm Evaluation for Resistance to Insect Pests

Germplasm of fruit, vegetable and ornamental crops, introduced and being maintained at the institute, was evaluated to find resistance sources to major insect and mite pests. The resistance or susceptibility of genotypes was also correlated with different morphological and biochemical parameters.

Screening of pests from voucher samples collected from explorations and collections

Mango: Leaf and fruit samples from exploration missions to South Andaman Islands were screened for insect pest damage. Leaf damage up to 25% was observed in samples collected from Baratang Island. All accessions were free from stone weevil damage. Leaf and fruit samples from exploration missions to Middle and North Andaman Islands were screened for insect pest damage. Leaf damage up to 46.58 % (8-10%) was observed.

Jackfruit: Average leaf damage of 26.3% (Range-18.18-37.50%) was recorded from two voucher samples.

Jamun: Average leaf damage of 45.48% (Range 37.35-56.66 %) was recorded from 3 voucher samples.

Dharwar and Kittur

Jamun: 16 Leaf and fruit samples from Dharwar and Kittur recorded leaf damage of 42.38 % (4.16%-90.9%).

Kaveripattanam and Paiyur, Tamil Nadu

8 Leaf and fruit samples from Kaveripattanam and Paiyur were screened for insect pest damage. Average Defoliator damage was 44.73% (26.40-82.91%) and Leaf folder damage was 91.62% (33-100%).

Khanapur, Belgaum

Leaf and fruit samples (32) from Belgaum, Khanapur and border areas of Maharashtra were screened for insect pest damage. Damage by leaf gall (14.28 %) caused by Psyllids was observed.

Monitoring and recording of pest load on domesticated, under utilised fruit species and medicinal and aromatic plants Under utilised fruit species: In Flacourtia montana, larvae and pupae of leaf feeder Phalanta phalanta was observed assuming serious pest status.

In Aegel marmelos

Domesticated fruit species: In Baccaurea courtallensis leaves damaged by mirid bugs, Helopeltis sp. On Aegle marmelos, leaf feeder beetle, Clitia picta causing serious leaf damge was observed.

Other pest species viz., citrus butterfly, Papilio demoleus, (0.5 larvae/shoot), and an ash weevil (Mylloceros sp) (22.5 -34.6% with mean of 28.55% leaf damage) were recorded. On Psidium spp., sporadic incidence of green scale (Coccus viridis), Bag worms, Spiraling whitefly (Aleurodicus disperses), Termites and Hairy caterpillar (Euproctis sp) were recorded. Besides these, activity of honey bees (Apis dorsata, and A. florea) was also observed on flowers of both the Psidium species.

Medicinal and Ornamentals: In Amla, leaf folder, Gracillaria acidula causes 9.56-38.53% leaf damage. In Bird of Paradise, Lymantrid larvae, Mealy bug (Ferrisia verghata) and 60% parasitization by Anesius advena was recorded

Mango: Stone weevil damage on seedling population of polyembryonic varieties of mango viz., Muvandan, Vellaikulumban, Mylipilian, Nekkare, Kensington, Olour, Prior, Moreh varied from 0 (Nekkare, Mylipilian, Prior, Manipur, Moreh)- 92.15 % (Olour) with a mean16.98%.

Seedlings of seven polyembryonic varieties of mango were examined for the infestation of mango scale (Aulacaspis tubercularis) under net house conditions. Variety Moreh was initially free from Scales but showed infestation after 3 months to the extent 25.40%. Bappakai recorded 82.26 % infestation. The damage caused in other varieties varied from 27.55 % (Kensington pride) to 48.49% (Vellaikulumban). Among the infested plant parts, lower surface of the leaf near the petiole was most affected and the damage varied between 30 and 48 %. After two weeks of spray with Imidacloprid @ 0.3 ml/L controlled the pest infestation in all varieties except Bappakai, where resurgence of the pest incidence was observed.

Pollinator diversity in mango germplasm: Recorded pollinators in nine varieties and one species. (Amrapali,Banganapalli, Kalepad, Prabhat Shankar, M. odorata, Langra, Himsagar, Lajjat Bakshi, Rumani, Totapuri). Apis florea, A. cerana, Eristalinus arvorum, Chrysomya megacephala, was found to be major pollinators across varieties in consecutive years. Exception observed is the foraging of rock bee, Apis dorsata (on Amrapali). Pollinators under protected cultivation: Introduction of honeybee colonies was not successful. Bees congregate to the corners of roof and die. However under net house, they were effective. Alternatively Calliphorid flies are being evaluated with watermelon.

Germplasm Conservation

1. Cryopreservation

Pollen cryobank conserves NGD of important horticultural crops. Long term conserved pollen in the form of nuclear genetic diversity (NGD) of citrus, papaya, grape, mango, tomato, eggplant, onion, capsicum, rose, gladiolus, gerbera, carnation and RET species of medicinal plants are continued to be cryopreserved in liquid nitrogen. The pollen cryobank was maintained and managed by periodic replenishment of the cryogen, for maintaining a constant cryogenic temperature throughout the storage duration.

2 In- vitro Conservation

Protocols have been optimized for 23 horticultural crops, which are being conserved under normal and reduced culture conditions. Vitroplants of Jackfruit accessions have been successfully conserved for 4 years under standard culture conditions prior to first subculture. While more jackfruit accessions are accessed in vitro, 4 citrus accessions are maintained in vitro and conservation attempts has resulted in maintaining vitro plants under reduced culture conditions for 6 months.

3. Seed Conservation

Seed conservation is aimed at maintenance of high seed quality in terms of viability and vigour for various periods. High seed quality was maintained for 20 years in onion, radish, French bean, cluster bean, watermelon, pumpkin and capsicum; for 15 years in chilli, carrot, tomato, brinjal, long melon, amaranth, bottle gourd, spinach beet and muskmelon and 8 years in cucumber and marigold by using moisture impervious containers and storing them at sub-zero temperature (-200C). Seeds preserved well in polyethylene bags too at low temperature (50C), which can also be used for medium term storage (<10 years) without affecting the seed quality. High seedling vigour in terms of growth parameters was exhibited in low temperature stored seeds. Further, seedlings emerged from stored seeds were healthy and normal. There were no changes in plant characteristics on germination of stored seed germplasm. A low cost storage technique such as modified atmosphere packaging of seeds with nitrogen or carbon dioxide showed relatively higher seed storability. Seed viability was greater with nitrogen in ridge gourd, garden peas and bitter gourd and with carbon dioxide in Knol Khol, cabbage, French bean and cluster bean under ambient conditions.

4. Field conservation

The division has been maintaining several accessions of different fruit crops in Field Gene Bank viz., guava (70 accessions), ber (30), jamun (12), custard apple (14) and fig (6). Collection of medicinally important species made during the previous years are being maintained in the form of FGB simulating wild situations under captivity in the form of a plant biodiversity block maintaining over 100 species collected from the Western Ghat region.

Bt cotton

Bt cotton is a genetically modified organism (GMO) or genetically modified pest resistant plant cotton variety, which produces an insecticide to bollworm.

Description

Strains of the bacterium Bacillus thuringiensis produce over 200 different Bt toxins, each harmful to different insects. Most notably, Bt toxins are insecticidal to the larvae of moths and butterflies, beetles, cotton bollworms and ghtu flies but are harmless to other forms of life. The gene coding for Bt toxin has been inserted into cotton as a transgene, causing it to produce this natural insecticide in its tissues. In many regions, the main pests in commercial cotton are lepidopteran larvae, which are killed by the Bt protein in the genetically modified cotton they eat. This eliminates the need to use large amounts of broad-spectrum insecticides to kill lepidopteran pests (some of which have developed pyrethroid resistance). This spares natural insect predators in the farm ecology and further contributes to noninsecticide pest management. Bt cotton is ineffective against many cotton pests such as plant bugs, stink bugs, and aphids; depending on circumstances it may be desirable to use insecticides in prevention. A 2006 study done by Cornell researchers, the Center for Chinese Agricultural Policy and the Chinese Academy of Science on Bt cotton farming in China found that after seven years these secondary pests that were normally controlled by pesticide had increased, necessitating the use of pesticides at similar levels to non-Bt cotton and causing less profit for farmers because of the extra expense of GM seeds.

Mechanism

Bt cotton was created through the addition of genes encoding toxin crystals in the Cry group of endotoxin. When insects attack and eat the cotton plant the Cry toxins are dissolved due to the high pH level of the insect's stomach. The dissolved and activated Cry molecules bond to cadherin-like proteins on cells comprising the brush border molecules. The epithelium of the brush border membranes separates the body cavity from the gut while allowing access for nutrients. The Cry toxin molecules attach themselves to specific locations on the cadherin-like proteins present on the epithelial cells of the midge and ion channels are formed which allow the flow of potassium. Regulation of potassium concentration is essential and, if left unchecked, causes death of cells. Due to the formation of Cry ion channels sufficient regulation of potassium ions is lost and results in the death of epithelial cells. The death of such cells creates gaps in the brush border membrane.

Bt cotton was first approved for field trials in the United States in 1993, and first approved commercial use in the United States in 1995. Bt cotton was approved by the Chinese government in 1997.

In 2002, a joint venture between Monsanto and Mahyco introduced Bt cotton to India. In 2011, India grew the largest GM cotton crop at 10.6 million hectares. The U.S.

GM cotton crop was 4.0 million hectares, the second largest area in the world, followed by China with 3.9 million hectares and Pakistan with 2.6 million hectares. By 2014, 96% of cotton grown in the United States was genetically modified and 95% of cotton grown in India was GM. India is the largest producer of cotton, and GM cotton, as of 2014.

Advantages

Bt cotton has several advantages over non-Bt cotton. The important advantages of Bt cotton are briefly :

- Increases yield of cotton due to effective control of three types of bollworms, viz. American, Spotted and Pink bollworms.
- Insects belonged to Lepidoptera (Bollworms) are sensitive to crystalline endotoxic protein produced by Bt gene which in turn protects cotton from bollworms.
- Reduction in insecticide use in the cultivation of Bt cotton in which bollworms are major pests.
- Potential reduction in the cost of cultivation (depending on seed cost versus insecticide costs).
- Reduction in es and predators which help in controlling the bollworms by feeding on larvae and eggs of bollworm.
- No health hazards due to rare use of insecticides (particularly who is engaged in spraying of insecticides).

The main selling points of Bt cotton are the reductions in pesticides to be sprayed on a crop and the ecological benefits which stem from that. China first planted Bt cotton in 1997 specifically in response to an outbreak of cotton bollworm, Helicoverpa armigera, that farmers were struggling to control with conventional pesticides. Similarly in India and the US, Bt cotton initially alleviated the issues with pests whilst increasing yields and delivering higher profits for farmers.

Studies showed that the lower levels of pesticide being sprayed on the cotton crops promoted biodiversity by allowing non-target species like ladybirds, lacewings and spiders to become more abundant. Likewise it was found that integrated pest management strategies (IPM) were becoming more effective due to the lower levels of pesticide encouraging the growth of natural enemy populations.

Disadvantages

Bt cotton has some limitations

- High cost of Bt cotton seeds as compared to non Bt cotton seeds.
- Effectiveness up to 120 days, after that the toxin producing efficiency of the Bt gene drastically reduces.
- Ineffective against sucking pests like jassids, aphids, whitefly etc.

In India

Bt cotton is supplied in India's Maharashatra by the agri-biotechnology company Mahyco, which distributes it. The use of Bt cotton in India has grown exponentially since its introduction in 2002.Recently India has become the number one global exporter of cotton and the second largest cotton producer in the world. India has bred Bt-cotton varieties such as Bikaneri Nerma and hybrids such as NHH-44.

Socio-economic surveys confirm that Bt cotton continues to deliver significant and multiple agronomic, economic, environmental and welfare benefits to Indian farmers and society including halved insecticide requirements and a doubling of yields. India's success has been subject to scrutiny. Monsanto's seeds are expensive and lose vigour after one generation, prompting the Indian Council of Agricultural Research to develop a cheaper Bt cotton variety with seeds that could be reused. The cotton incorporated the cry1Ac gene from the soil bacterium Bacillus thuringiensis (Bt), making the cotton toxic to bollworms. This variety showed poor yield, was removed within a year, and contained a DNA sequence owned by Monsanto, prompting an investigation. In parts of India cases of acquired resistance against Bt cotton have occurred. Monsanto has admitted that the pink bollworm is resistant to first generation transgenic Bt cotton that expresses the single Bt gene (Cry1Ac).

The state of Maharashtra banned the sale and distribution of Bt cotton in 2012, to promote local Indian seeds, which demand less water, fertilizers and pesticide input, but lifted the ban in 2013.Punjab Agricultural University(PAU) has successfully developed the country's first Bt cotton varieties. ICAR has identified three varieties, namely PAU Bt 1, F1861 and RS 2013, for cultivation in Punjab, Haryana, Rajasthan. It is a cheaper alternative to Bt cotton hybrid seed.

Bt-resistant pests

After the introduction of Bt cotton in northern China, non-target pests such as mirid bugs (Heteroptera: Miridae) became more abundant, because less pesticides were sprayed. In 2013, a second issue being seen across the world, was the development of Bt resistant pests limiting the usefulness of Bt crops.

Main drivers for the widespread resistance in India and China included the high proportion of Bt cotton being planted, 90% and 95% respectively in 2011, and few refuge areas. Refuge areas of non-Bt crops limit resistance development in targeted pests. The US Environmental Protection Agency requires farmers to have refuge areas of 20%-50% non-Bt crops within 0.8 km of their Bt fields. Such requirements were not seen in China, where instead farmers relied on natural refuge areas to decrease resistance. In 2009, a novel solution to the resistance problem was trialed in Arizona, when sterile male pink bollworms (Pectinophora gossypiella) were released into populations of their wild Bt-resistant counterparts. The hypothesis was that sterile males mating with the few surviving females, who had developed resistance, would lead to a decrease in pests in the following generation. There was a dramatic reduction in pink bollworms, with only two pink bollworm larvae being found by the third year of the study.

Controversies

In India, Bt cotton has been enveloped in controversies due to its supposed failure to reduce the need for pesticides and increase yield. The link between the introduction of Bt cotton to India and a surge in farmer suicides has been refuted by other studies, with decreased farmer suicides since Bt cotton was introduced. Bt cotton accounts for 93% of cotton grown in India.

In the USA

In Hawaii, growing GMO cotton has been prohibited since 2013. Hybridization with the wild cotton species Gossypium tomentosum may be possible. Transgenic cotton is also banned in some parts of Florida.

Genetically modified brinjal

The genetically modified brinjal is a suite of transgenicbrinjals (also known as an eggplant or aubergine) created by inserting a crystal protein gene (Cry1Ac) from the soil bacteriumBacillus thuringiensis into the genome of various brinjal cultivars. The insertion of the gene, along with other genetic elements such as promoters, terminators and an antibiotic resistance marker gene into the brinjal plant is accomplished using Agrobacterium-mediated genetic transformation. The Bt brinjal has been developed to give resistance against lepidopteron insects, in particular the Brinjal Fruit and Shoot Borer (Leucinodes orbonalis)(FSB). Mahyco, an Indian seed company based in Jalna, Maharashtra, has developed the Bt brinjal.

The genetically modified brinjal event is termed Event EE 1, and Mahyco have also applied for approval of two brinjal hybrids. The Event EE 1 was introgressed by plant breeding into various local varieties by University of Agricultural Sciences, Dharwad and Tamil Nadu Agricultural University, Coimbatore. Some of the cultivars of brinjal include: Malpur local, Manjari gota, Kudachi local, Udupi local, 112 GO, and Pabkavi local. It was approved for commercialization in India in 2009, but - after an apparent public outcry and rounds of debates in which representatives from Mahyco, the scientific community, and NGO's spoke on the topic - the then Indian Environment Minister, Jairam Ramesh, facilitated a moratorium on its release until further, unspecified, tests were conducted. Bt brinjal was approved for commercial release in Bangladesh in 2013.

GM brinjal isn't legal in India

Arjun Saini, director general of horticulture, Haryana, said "The government is very serious about this issue. Environment department will take further measures like investigation and FIR as per the law." Dhirendra Khadgata, deputy commissioner,

Fatehabad, said no FIR has been registered. The Genetic Engineering Appraisal Committee (GEAC) under the Union environment ministry, however, did not comment on whether possible contamination of GM brinjal is being investigated elsewhere in the country.

Earlier this month, farmers in Haryana and experts had expressed fears of widespread contamination of GM brinjal in the state with the farmer accused of cultivating Bt brinjal in Fatehabad district admitting that he had been growing and selling the GM produce since 2017. Farm activists demanded that other brinjal fields in Haryana be investigated for contamination. "There are several things to be resolved yet. For one, the source of seedling supply has to be zeroed in on by Haryana government. That should not be very difficult given four farmers that we are aware of, grew this unapproved GM brinjal, and because there would only be so many nurseries, in Haryana or Punjab," said Kavitha Kuriganti, convenor of Alliance for Sustainable & Holistic Agriculture.

"A village-wise survey should be undertaken of all brinjal farmers and their past three years' cultivation experience. Meanwhile, GEAC has to instruct states like Punjab to take up a survey too," she added. Earlier this week, Saini had claimed citing the results received from National Bureau of Plant Genetic Resources (NBPGR), that "the samples have tested negative for the CRY 1AC gene and EE 1 event, so these may not have been contaminated by the variety being grown in Bangladesh".

"They have tested positive for other gene markers which will have to be investigated by the Haryana Agricultural University in Hisar," he said. But it's still not clear which GM version may have been leaked.

A few days ago, Aruna Rodrigues, the lead petitioner in the Supreme Court against genetically modified crops, wrote to Anil Kumar Jain, chairman of GEAC calling for event-specific testing of the samples. [An event is the insertion of a particular gene] .

Disadvantages and Advantages of Genetically Modified Crops

Nearly most of the crops that are grown in the US have been touched by science. Having genes that have been manipulated, genetically modified crops are becoming more and more popular in the country, with people liking the fact that they provide certain perks, such as larger yields, insect resistance and faster growing traits. However, they also do not come with criticism, where many people believe that research on the technology is not sufficient. In order to get a better understanding of genetically modified crops, it is best to take a look at their pros and cons.

List of Pros of Genetically Modified Crops

1. They are a technology that has already been going on for many years.- Genetic modification and engineering is nothing new, where people have been already manipulating crops and food for many years, using methods that range from classical selection to fermentation. It is just the latest form of biotechnology that is using the most precise method known. Basically, creating these crops is different from the traditional methods of plant breeding, as it transfers genes from one species to another, crossing biological barriers.

2. They are mostly believed to be safe.-Proponents say that there are actually no inherent differences between food items produced from natural crops and those from genetically modified crops, similarly containing DNA that comprised of nucleotides. This technology works by moving a piece of DNA from one plant breed into another, without introducing a foreign substance and with the new gene merely prompting the modified crop to show a desired trait. Also companies wishing to release a product or seed from a genetically modified crop are required the safety of these things. And if a certain product was created from an organism that contains a known allergen, it is required to undergo some tests for safety.

3. They have improved quality and taste, as well as increased nutrients.- Genetic modification in crops allows flavors of its products to be improved. For example, pepper can become spicier, corn can become sweeter and bitter gourd would become palatable without losing their original nutrients. Better yet, this technology makes it possible to enhance the vitamin and mineral content of food, providing benefits to consumers. This advantage is especially important to people in developing countries, who do not always have access to essential nutritional resources.

4. They help ease the environmental impact by human activities.-With the way it is practiced today, agriculture sometimes damages the environment, like any other human activity. Now, genetically modified crops can help ease such negative impact by creating plants that are insect resistant, allowing farmers to dramatically reduce the use of insecticides. Also, the technology can create next-generation seeds that will allow farmers to maintain high yields, while using fewer amounts of chemical fertilizer and water. While critics point out the creation of super pests and weeds, which are highly possible, these dangers can be prevented and are manageable.

5. They can help reduce world hunger.-With genetic modification, even the smallest subsistence farmers will be able to produce larger, more reliable and profitable crops for the market, as well as able to grow more food for their families. This means that we will be able to directly address the problem of hunger and malnutrition among poor consumers, especially in the developing countries. In the long run, genetically modified crops are considered to be a solution to ensuring that food production can keep pace with the growing population, which is expected to double 12 billion by 2050. The rate of growth of food production has declined in the past decade, with the latest round of increase driven by Green Revolution methods, such as using high-yielding hybrid seeds, extensive amounts of fertilizer and pesticides. However, these technologies do not have the capability to support food production growth for the coming decades without resorting to things that can severely damage the environment. Now, GM crops are believed to be able to support such growth.

List of Cons of Genetically Modified Crops

1. They are not entirely believed to be safe.-Truth is, there are also a number of independent studies of the health effects of genetically modified crops that do not have confidence in their safety. In an experiment in Scotland, for example, mice that were fed with modified potatoes containing lectin fared poorly and even suffered damage to their internal organs. While supporters have attacked the study, it entails the need for more research at the very least.

2. They contribute to the increased occurrence of allergic reactions. -Many studies show that consuming genetically modified crops has increased people's risk of acquiring allergies that are food-related, with symptoms including fatigue, rashes, headache and nausea.

3. They alter the way nature works.-Genetically engineered crops are believed to do wide-reaching damage to the environment. For example, insect-resistant plants might harm insects that are not their intended target, such as monarch butterflies. Also, target insects could develop resistance to these crops, requiring farmers to resort to using more aggressive measures to control the problem, such as extensive use of chemical sprays. As previously implied, the technology could also allow the transfer of genes from one crop to another, creating "super weeds", which will be immune to common control methods. Moreover, viral genes added to crops to confer resistance might be transferred to other viral pathogens, which can lead to new and more virulent virus strains.

4. They require labeling.-Food labeling can incite fear, causing the public not to accept food items made from genetically modified crops. As you can see, consumers might object to consuming of GM foods on the basis of health, ethical and religious concerns. Lack of evidence that proves that GM products are unsafe must not be taken as proof that they are actually safe.

5. They only have little to no economic value.-While genetically modified crops might be resistant to herbicides and pesticides, as well as contain more nutrients, they actually contribute little to the farmers' profits. They take more time to mature and more effort to grow compared to natural crops.

Now, do you think that genetically modified crops are good for society or not? Do their benefits outdo the risks? By carefully weighing the pros and cons listed above, you can be a better judge.

Development

Mahyco licensed and used the cry1Ac gene obtained from Monsanto and two supporting genes (nptII and aad). The cry1Ac gene is under the transcriptional control of an enhanced cauliflower mosaic virus 35S (CaMV35S) promoter, which ensures the gene is expressed in all the brinjal's tissue throughout its complete life cycle. NptII and aad are selectable marker genes, nptII is used to identify transgenic plants from non-transgenic, and aad is used to identify the transformed bacteria used during the development of the construct. Aad contains a bacterial promoter and is not expressed in the Bt brinjal. The completed construct was inserted into young cotyledons from the brinjal plants using an Agrobacterium-mediated technique. Agrobacterium naturally inserts DNA into plants from its Ti plasmid, and scientists use this to insert genes of interest into various plants. The transformed plants were regenerated and analyzed for the presence of the gene through Southern blotting. The plants' progeny were also analyzed to identify lines segregating in a Mendelian fashion.

Effective against pests

When fruit and shoot borer larvae feed on Bt brinjal plants, they ingest the Bt protein Cry1Ac along with plant tissue. In the insect gut, which is alkaline with a pH >9.5, the protein is solubilized and activated by gut proteases. The Bt protein binds to specific receptor proteins present in the insect membrane, resulting in pore formation in the membranes. This leads to disruption of digestive processes, paralysis, and subsequent death of the fruit and shoot borer larvae.

Attempted commercialization in India

The first agreement to develop Bt Brinjal was signed in 2005 between India's leading seed company, Maharashtra Hybrid Seeds Company, better known as Mahyco, and two agricultural universities - University of Agricultural Sciences, Dharwad (UAS) and Tamil Nadu Agricultural University (TNAU) in Coimbatore.

An expert committee (EC-I) was set up in 2006 to examine the biosafety data presented by Mahyco. They concluded that while the current data demonstrated that Bt brinjal was safe and equivalent to its non-Bt counterpart, more studies were required to re-affirm the findings and further trials were needed to ascertain the benefits from Bt brinjal with respect to existing methods for pest management and pesticide reduction. They recommended that large scale trials be allowed to go ahead. In 2009, a second expert committee (EC-II) examined the data from these trials. They concluded that adequate safety tests had been performed, stating that "the benefits of Bt brinjal event EE-I developed by M/s Mahyco far outweigh the perceived and projected risks", and advised the Genetic Engineering Appraisal Committee (GEAC) to recommend commercialization of the Bt brinjal.

The GEAC cleared Bt brinjal for commercialization on 14 October 2009. Following concerns raised by some scientists, farmers and anti-GMO activists, the government of India officially announced on 9 February 2010 that it needed more time before releasing Bt brinjal, with Indian Environment Minister Jairam Ramesh saying that there is no overriding urgency to introduce Bt brinjal in India. On 17 February 2010, Jairam Ramesh reiterated that the Centre had only imposed a moratorium on the release of transgenic brinjal hybrid, and not a permanent ban, saying that "until we arrive at a political, scientific and societal consensus, this moratorium will remain". Companies with any seeds of Bt brinjal will have to register the details with the government, and the National Bureau of Plant Genetic Resources (NBPGR) was made responsible for storage of all the Bt brinjal seeds in India. Independent testing labs are currently being set up.

An irregularity was also brought to the notice of the Karnataka Biodiversity Board by Environment Support Group, a charitable trust in Bengaluru, in February 2010. It found that agencies accessed at least 10 brinjal varieties from Karnataka and Tamil Nadu without seeking prior consent of the National Biodiversity Authority and state biodiversity boards. Mahyco became India's first commercial entity to be accused of bio-piracy, or misappropriation, of local germplasm. In October 2013, the Indian High Court was pursuing criminal proceedings against senior officials of Mahyco-Monsanto.

Commercial Cultivation in Bangladesh

On 30 October 2013 with approvals from the ministries of Environment and Forests (MoEF) and Agriculture (MoA), the Bangladesh Agricultural Research Institute (BARI) received permission to release four varieties of Bt brinjal in time for the 2013–2014 growing season: Bt Uttara, Bt Kajla, Bt Nayantara, and Bt ISD006. The Bt varieties underwent seven years of field and greenhouse trials in various environmental and geographic locations in Bangladesh Saplings were distributed to 20 farmers in January 2014 The Guardian spoke to 19 of 20 farmers growing the Bt brinjal crop in 2014 and established that it has so far had mixed results. In September 2016 the Indian Business Standard quoted a Director of the Bangladesh Department of Environment who said that production results had been very good from the 200-300 farmers who had grown the bt brinjal since 2013 but it was too early to judge whether there had been any contamination of wild brinjal.

Genetically modified food controversies

Genetically modified food controversies are disputes over the use of foods and other goods derived from genetically modified crops instead of conventional crops, and other uses of genetic engineering in food production. The disputes involve consumers, farmers, biotechnology companies, governmental regulators, non-governmental organizations, and scientists. The key areas of controversy related to genetically modified food (GM food or GMO food) are whether such food should be labeled, the role of government regulators, the objectivity of scientific research and publication, the effect of genetically modified crops on health and the environment, the effect on pesticide resistance, the impact of such crops for farmers, and the role of the crops in feeding the world population. In addition, products derived from GMO organisms play a role in the production of ethanol fuels and pharmaceuticals.

Specific concerns include mixing of genetically modified and non-genetically modified products in the food supply, effects of GMOs on the environment, the rigor of the regulatory process, and consolidation of control of the food supply in companies that make and sell GMOs.Advocacy groups such as the Center for Food Safety, Organic

Consumers Association, Union of Concerned Scientists, and Greenpeace say risks have not been adequately identified and managed, and they have questioned the objectivity of regulatory authorities.

The safety assessment of genetically engineered food products by regulatory bodies starts with an evaluation of whether or not the food is substantially equivalent to non-genetically engineered counterparts that are already deemed fit for human consumption. No reports of ill effects have been documented in the human population from genetically modified food. There is a scientific consensus that currently available food derived from GM crops poses no greater risk to human health than conventional food, but that each GM food needs to be tested on a case-by-case basis before introduction. Nonetheless, members of the public are much less likely than scientists to perceive GM foods as safe. The legal and regulatory status of GM foods varies by country, with some nations banning or restricting them, and others permitting them with widely differing degrees of regulation.

Public perception

Consumer concerns about food quality first became prominent long before the advent of GM foods in the 1990s. Upton Sinclair's novel The Jungle led to the 1906 Pure Food and Drug Act, the first major US legislation on the subject. This began an enduring concern over the purity and later "naturalness" of food that evolved from a single focus on sanitation to include others on added ingredients such as preservatives, flavors and sweeteners, residues such as pesticides, the rise of organic food as a category and, finally, concerns over GM food. Some consumers, including many in the US, came to see GM food as "unnatural", with various negative associations and fears (a reverse halo effect).

Specific perceptions include a view of genetic engineering as meddling with naturally evolved biological processes, and one that science has limitations on its comprehension of potential negative ramifications. An opposing perception is that genetic engineering is itself an evolution of traditional selective breeding, and that the weight of current evidence suggests current GM foods are identical to conventional foods in nutritional value and effects on health.

Surveys indicate widespread concern among consumers that eating genetically modified food is harmful, that biotechnology is risky, that more information is needed and that consumers need control over whether to take such risks. A diffuse sense that social and technological change is accelerating, and that people cannot affect this context of change, becomes focused when such changes affect food. Leaders in driving public perception of the harms of such food in the media include Jeffrey M. Smith, Dr. Oz, Oprah, and Bill Maher; organizations include Organic Consumers Association, Greenpeace (especially with regard to Golden rice) and Union of Concerned Scientists. In the United States support or opposition or skepticism about GMO food is not divided by traditional partisan (liberal/conservative) lines, but young adults are more likely to have negative opinions on genetically modified food than older adults. Religious groups have raised concerns over whether genetically modified food will remain kosher or halal. In 2001, no such foods had been designated as unacceptable by Orthodox rabbis or Muslim leaders. However, some Jewish groups dispute this designation.

Food writer Michael Pollan does not oppose eating genetically modified foods, but supports mandatory labeling of GM foods and has criticized the pesticide-heavy monoculture farming enabled by certain GM crops, such as glycosphate-tolerant ("Roundup-ready") corn and soybeans. He has also expressed concerns about biotechnology companies holding the intellectual property of the foods people depend on, and about the effects of the growing corporatization of large-scale agriculture. To address these problems, Pollan has brought up the idea of open sourcing GM foods. The idea has since been adopted to varying degrees by companies like Syngenta, and is being promoted by organizations such as the New America Foundation. Some organizations, like The BioBricks Foundation, have already worked out open-source licenses that could prove useful in this endeavour.

Public Relations Campaigns and Protests

In May 2012, a group called "Take the Flour Back" led by Gerald Miles protested plans by a group from Rothamsted Experimental Station, based in Harpenden, Hertfordshire, England, to conduct an experimental trial wheat genetically modified to repel aphids. The researchers, led by John Pickett, wrote a letter to the group in early May 2012, asking them to call off their protest, aimed for 27 May 2012. Group member Lucy Harrap said that the group was concerned about spread of the crops into nature, and cited examples of outcomes in the United States and Canada. Rothamsted Research and Sense About Science ran question and answer sessions about such a potential.

The March Against Monsanto is an international grassroots movement and protest against Monsanto corporation, a producer of genetically modified organism (GMOs) and Roundup, a glyphosate-based herbicide. The movement was founded by Tami Canal in response to the failure of California Proposition 37, a ballot initiative which would have required labeling food products made from GMOs. Advocates support mandatory labeling laws for food made from GMOs . The initial march took place on May 25, 2013. The number of protesters who took part is uncertain; figures of "hundreds of thousands" and the organizers' estimate of "two million" were variously cited. Events took place in between 330 and 436 cities around the world, mostly in the United States. Many protests occurred in Southern California, and some participants carried signs expressing support for mandatory labeling of GMOs that read "Label GMOs, It's Our Right to Know", and "Real Food 4 Real People". Canal said that the movement would continue its "anti-GMO cause" beyond the initial event. Further marches occurred in October 2013 and in May 2014 and 2015. The protests were reported by news outlets including ABC News, the

Associated Press,The Washington Post,The Los Angeles Times,USA Today, and CNN (in the United States), and Russia Today and The Guardian (outside the United States).

Monsanto said that it respected people's rights to express their opinion on the topic, but maintained that its seeds improved agriculture by helping farmers produce more from their land while conserving resources, such as water and energy. The company reiterated that genetically modified foods were safe and improved crop yields. Similar sentiments were expressed by the Hawaii Crop Improvement Association, of which Monsanto is a member. In July 2013, the agricultural biotechnology industry launched a GMO transparency initiative called GMO Answers to address consumers' questions about GM foods in the U.S. food supply. GMO Answers' resources included conventional and organic farmers, agribusiness experts, scientists, academics, medical doctors and nutritionists, and "company experts" from founding members of the Council for Biotechnology Information, which funds the initiative. Founding members include BASF, Bayer CropScience, Dow AgroSciences, DuPont, Monsanto Company and Syngenta. In October 2013, a group called The European Network of Scientists for Social and Environmental Responsibility (ENSSER), posted a statement claiming that there is no scientific consensus on the safety of GMOs, which was signed by about 200 scientists in various fields in its first week. On January 25, 2015, their statement was formally published as a whitepaper by Environmental Sciences Europe:

Direct action

Earth Liberation Front, Greenpeace and others have disrupted GMO research around the world. Within the UK and other European countries, as of 2014 80 crop trials by academic or governmental research institutes had been destroyed by protesters.In some cases, threats and violence against people or property were carried out.In 1999, activists burned the biotech lab of Michigan State University, destroying the results of years of work and property worth $400,000.

In 1987, the ice-minus strain of P. syringae became the first genetically modified organism (GMO) to be released into the environment when a strawberry field in California was sprayed with the bacteria. This was followed by the spraying of a crop of potato seedlings.The plants in both test fields were uprooted by activist groups, but were re-planted the next day. In 2011, Greenpeace paid reparations when its members broke into the premises of an Australian scientific research organization, CSIRO, and destroyed a genetically modified wheat plot. The sentencing judge accused Greenpeace of cynically using junior members to avoid risking their own freedom. The offenders were given 9-month suspended sentences.

On August 8, 2013 protesters uprooted an experimental plot of golden rice in the Philippines. British author, journalist, and environmental activist Mark Lynas reported in Slate that the vandalism was carried out by a group led by the extreme-left KMP, to the dismay of other protesters. Golden rice is designed prevent vitamin A deficiency which, according to Helen Keller International, blinds or kills hundreds of thousands of children annually in developing countries.

Response to anti-GMO sentiment

In 2017, two documentaries were released which countered the growing anti-GMO sentiment among the public. These included Food Evolution and Science Moms. Per the Science Moms director, the film "focuses on providing a science and evidence-based counter-narrative to the pseudoscience-based parenting narrative that has cropped up in recent years".

Conspiracy theories

There are various conspiracy theories related to the production and sale of genetically modified crops and genetically modified food that have been identified by some commentators such as Michael Shermer.Generally, these conspiracy theories posit that GMOs are being knowingly and maliciously introduced into the food supply either as a means to unduly enrich agribusinesses or as a means to poison or pacify the population.

A work seeking to explore risk perception over GMOs in Turkey identified a belief among the conservative political and religious figures who were opposed to GMOs that GMOs were "a conspiracy by Jewish Multinational Companies and Israel for world domination."Additionally, a Latvian study showed that a segment of the population believed that GMOs were part of a greater conspiracy theory to poison the population of the country.

Bt maize

One of the first incidents occurred in 1999, when Nature published a paper on potential toxic effects of Bt maize on butterflies. The paper produced a public uproar and demonstrations, however by 2001 multiple follow-up studies had concluded that "the most common types of Bt maize pollen are not toxic to monarchlarvae in concentrations the insects would encounter in the fields" and that they had "brought that particular question to a close".

Concerned scientists began to patrol the scientific literature and react strongly, both publicly and privately, to discredit conclusions they view as flawed in order to prevent unjustified public outcry and regulatory action. A 2013 Scientific American article noted that a "tiny minority" of biologists have published concerns about GM food, and said that scientists who support the use of GMOs in food production are often overly dismissive of them.

Restrictive end-user agreements

Prior to 2010, scientists wishing to conduct research on commercial GM plants or seeds were unable to do so, because of restrictive end-user agreements. Cornell University's Elson Shields was the spokesperson for one group of scientists who opposed such restrictions. The group submitted a statement to the United States Environmental Protection Agency (EPA) in 2009 protesting that "as a result of restrictive access, no truly independent research can be legally conducted on many critical questions regarding the technology".

A 2009 Scientific American editorial quoted a scientist who said that several studies that were initially approved by seed companies were blocked from publication when they returned "unflattering" results. While favoring protection of intellectual property rights, the editors called for the restrictions to be lifted and for the EPA to require, as a condition of approval, that independent researchers have unfettered access to genetically modified products for research. In December 2009, the American Seed Trade Association agreed to "allow public researchers greater freedom to study the effects of GM food crops". The companies signed blanket agreements permitting such research. This agreement left many scientists optimistic about the future; other scientists still express concern as to whether this agreement has the ability to "alter what has been a research environment rife with obstruction and suspicion". Monsanto previously had research agreements (i.e., Academic Research Licenses) with approximately 100 universities that allowed for university scientists to conduct research on their GM products with no oversight.

Alleged data manipulation

A University of Naples investigation suggested that images in eight papers on animals were intentionally altered and/or misused. The leader of the research group, Federico Infascelli, rejected the claim. The research concluded that mother goats fed GM soybean meal secreted fragments of the foreign gene in their milk. In December 2015 one of the papers was retracted for "self-plagiarism", although the journal noted that the results remained valid.A second paper was retracted in March 2016 after The University of Naples concluded that "multiple heterogeneities were likely attributable to digital manipulation, raising serious doubts on the reliability of the findings".

Health

There is a scientific consensus that currently available food derived from GM crops poses no greater risk to human health than conventional food, but that each GM food needs to be tested on a case-by-case basis before introduction. Nonetheless, members of the public are much less likely than scientists to perceive GM foods as safe. The legal and regulatory status of GM foods varies by country, with some nations banning or restricting them, and others permitting them with widely differing degrees of regulation.

The ENTRANSFOOD project was a European Commission-funded scientist group chartered to set a research program to address public concerns about the safety and value of agricultural biotechnology. It concluded that "the combination of existing test methods provides a sound test-regime to assess the safety of GM crops." In 2010, the European Commission Directorate-General for Research and Innovation reported that "The main conclusion to be drawn from the efforts of more than 130 research projects, covering a period of more than 25 years of involving more than 500 independent research groups, is that biotechnology, and in particular GMOs, are not per se more risky than e.g. conventional plant breeding technologies."

Comparison of conventional plant breeding with transgenic and cisgenic genetic modification. Consensus among scientists and regulators pointed to the need for improved testing technologies and protocols.Transgenic and cisgenic organisms are treated similarly when assessed. However, in 2012 the European Food Safety Authority (EFSA) GMO Panel said that "novel hazards" could be associated with transgenic strains. In a 2016 review, Domingo concluded that studies in recent years had established that GM soybeans, rice, corn, and wheat do not differ from the corresponding conventional crops in terms of short-term human health effects, but recommended that further studies of long-term effects be conducted.

Substantial equivalence

Most conventional agricultural products are the products of genetic manipulation via traditional cross-breeding and hybridization. Governments manage the marketing and release of GM foods on a case-by-case basis. Countries differ in their risk assessments and regulations. Marked differences distinguish the US from Europe. Crops not intended as foods are generally not reviewed for food safety. GM foods are not tested in humans before marketing because they are not a single chemical, nor are they intended to be ingested using specific doses and intervals, which complicate clinical study design. Regulators examine the genetic modification, related protein products and any changes that those proteins make to the food.

Regulators check that GM foods are "substantially equivalent" to their conventional counterparts, to detect any negative unintended consequences. New protein(s) that differ from conventional food proteins or anomalies that arise in the substantial equivalence comparison require further toxicological analysis.

"The World Health Organization, the American Medical Association, the U.S. National Academy of Sciences, the British Royal Society, and every other respected organization that has examined the evidence has come to the same conclusion: consuming foods containing ingredients derived from GM crops is no riskier than consuming the same foods containing ingredients from crop plants modified by conventional plant improvement techniques." –American Association for the Advancement of Science

In 1999, Andrew Chesson of the Rowett Research Institute warned that substantial equivalence testing "could be flawed in some cases" and that current safety tests could allow harmful substances to enter the human food supply. The same year Millstone, Brunner and Mayer argued that the standard was a pseudo-scientific product of politics and lobbying that was created to reassure consumers and aid biotechnology companies to reduce the time and cost of safety testing. They suggested that GM foods have extensive biological, toxicological and immunological tests and that substantial equivalence should be abandoned. This commentary was criticized for misrepresenting history, for distorting existing data and poor logic. Kuiper claimed that it oversimplified safety assessments and that equivalence testing involves more than chemical tests, possibly including toxicity testing. Keler and Lappe supported Congressional legislation to replace the substantial equivalence standard with safety studies.In a 2016 review, Domingo criticized the use of the "substantial equivalence" concept as a measure of the safety of GM crops.

Kuiper examined this process further in 2002, finding that substantial equivalence does not measure absolute risks, but instead identifies differences between new and existing products. He claimed that characterizing differences is properly a starting point for a safety assessment and "the concept of substantial equivalence is an adequate tool in order to identify safety issues related to genetically modified products that have a traditional counterpart". Kuiper noted practical difficulties in applying this standard, including the fact that traditional foods contain many toxic or carcinogenic chemicals and that existing diets were never proven to be safe.

This lack of knowledge re conventional food means that modified foods may differ in anti-nutrients and natural toxins that have never been identified in the original plant, possibly allowing harmful changes to be missed. In turn, positive modifications may also be missed. For example, corn damaged by insects often contains high levels of fumonisins, carcinogenic toxins made by fungi that travel on insects' backs and that grow in the wounds of damaged corn. Studies show that most Bt corn has lower levels of fumonisins than conventional insect-damaged corn. Workshops and consultations organized by the OECD, WHO, and FAO have worked to acquire data and develop better understanding of conventional foods, for use in assessing GM foods.

A survey of publications comparing the intrinsic qualities of modified and conventional crop lines (examining genomes, proteomes and metabolomes) concluded that GM crops had less impact on gene expression or on protein and metabolite levels than the variability generated by conventional breeding.

In a 2013 review, Herman (Dow AgroSciences) and Price (FDA, retired) argued that transgenesis is less disruptive than traditional breeding techniques because the latter routinely involve more changes (mutations, deletions, insertions and rearrangements) than the relatively limited changes (often single gene) in genetic engineering. The FDA found that all of the 148 transgenic events that they evaluated to be substantially equivalent to their conventional counterparts, as have Japanese regulators for 189 submissions including combined-trait products. This equivalence was confirmed by more than 80 peer-reviewed publications. Hence, the authors argue, compositional equivalence studies uniquely required for GM food crops may no longer be justified on the basis of scientific uncertainty.

Allergenicity

A well-known risk of genetic modification is the introduction of an allergen. Allergen testing is routine for products intended for food, and passing those tests is part of the regulatory requirements. Organizations such as the European Green Party and Greenpeace emphasize this risk.A 2005 review of the results from allergen testing stated that "no biotech proteins in foods have been documented to cause allergic reactions". Regulatory authorities require that new modified foods be tested for allergenicity before they are marketed.

GMO proponents note that because of the safety testing requirements, the risk of introducing a plant variety with a new allergen or toxin is much smaller than from traditional breeding processes, which do not require such tests. Genetic engineering can have less impact on the expression of genomes or on protein and metabolite levels than conventional breeding or (non-directed) plant mutagenesis. Toxicologists note that "conventional food is not risk-free; allergies occur with many known and even new conventional foods. For example, the kiwi fruit was introduced into the U.S. and the European markets in the 1960s with no known human allergies; however, today there are people allergic to this fruit."

Genetic modification can also be used to remove allergens from foods, potentially reducing the risk of food allergies.A hypo-allergenic strain of soybean was tested in 2003 and shown to lack the major allergen that is found in the beans. A similar approach has been tried in ryegrass, which produces pollen that is a major cause of hay fever: here a fertile GM grass was produced that lacked the main pollen allergen, demonstrating that hypoallergenic grass is also possible.

The development of genetically modified products found to cause allergic reactions has been halted by the companies developing them before they were brought to market. In the early 1990s, Pioneer Hi-Bred attempted to improve the nutrition content of soybeans intended for animal feed by adding a gene from the Brazil nut. Because they knew that people have allergies to nuts, Pioneer ran in vitro and skin prick allergy tests. The tests showed that the transgenic soy was allergenic. Pioneer Hi-Bred therefore discontinued further development. In 2005, a pest-resistant field pea developed by the Australian Commonwealth Scientific and Industrial Research Organisation for use as a pasture crop was shown to cause an allergic reaction in mice. Work on this variety was immediately halted. These cases have been used as evidence that genetic modification can produce unexpected and dangerous changes in foods, and as evidence that safety tests effectively protect the food supply.

During the Starlink corn recalls in 2000, a variety of GM maize containing the Bacillus thuringiensis (Bt) protein Cry9C, was found contaminating corn products in U.S. supermarkets and restaurants. It was also found in Japan and South Korea. Starlink corn had only been approved for animal feed as the Cry9C protein lasts longer in the digestive system than other Bt proteins raising concerns about its potential allergenicity.In 2000, Taco Bell-branded taco shells sold in supermarkets were found to contain Starlink, resulting in a recall of those products, and eventually led to the recall of over 300 products. Sales of StarLink seed were discontinued and the registration for the Starlink varieties was voluntarily withdrawn by Aventis in October 2000. Aid sent by the United Nations and the United States to Central African nations was also found to be contaminated with StarLink corn and the aid was rejected.

The U.S. corn supply has been monitored for Starlink Bt proteins since 2001 and no positive samples have been found since 2004.In response, GeneWatch UK and Greenpeace set up the GM Contamination Register in 2005. During the recall, the United States Centers for Disease Control evaluated reports of allergic reactions to StarLink corn, and determined that no allergic reactions to the corn had occurred.

Horizontal gene Transfer

Horizontal gene transfer is the movement of genes from one organism to another in a manner other than reproduction. The risk of horizontal gene transfer between GMO plants and animals is very low and in most cases is expected to be lower than background rates. Two studies on the possible effects of feeding animals with genetically modified food found no residues of recombinant DNA or novel proteins in any organ or tissue samples. Studies found DNA from the M13 virus, Green fluorescent protein and RuBisCO genes in the blood and tissue of animals, and in 2012, a paper suggested that a specific microRNA from rice could be found at very low quantities in human and animal serum. Other studies however, found no or negligible transfer of plant microRNAs into the blood of humans or any of three model organisms. Another concern is that the antibiotic resistance gene commonly used as a genetic marker in transgenic crops could be transferred to harmful bacteria, creating resistant superbugs.

A 2004 study involving human volunteers examined whether the transgene from modified soy would transfer to bacteria that live in the human gut. As of 2012 it was the only human feeding study to have been conducted with GM food. The transgene was detected in three volunteers from a group of seven who had previously had their large intestines removed for medical reasons. As this gene transfer did not increase after the consumption of the modified soy, the researchers concluded that gene transfer did not occur. In volunteers with intact digestive tracts, the transgene did not survive. The antibiotic resistance genes used in genetic engineering are naturally found in many pathogens and antibiotics these genes confer resistance to are not widely prescribed.

Nutritional quality

Some plants are specifically genetically modified to be healthier than conventional crops. Golden rice was created to combat vitamin A deficiency by synthesizing beta carotene (which conventional rice does not).

Detoxification

One variety of cottonseed has been genetically modified to remove the toxin gossypol, so that it would be safe for humans to eat.

Environment

Genetically modified crops are planted in fields much like regular crops. There they interact directly with organisms that feed on the crops and indirectly with other organisms in the food chain. The pollen from the plants is distributed in the environment like that of any other crop. This distribution has led to concerns over the effects of GM crops on the environment. Potential effects include gene flow, pesticide resistance and greenhouse gas emissions.

Non-target Organisms

A major use of GM crops is in insect control through the expression of the cry (crystal delta-endotoxins) and Vip (vegetative insecticidal proteins) genes from Bacillus thuringiensis (Bt). Such toxins could affect other insects in addition to targeted pests such as the European corn borer. Bt proteins have been used as organic sprays for insect control in France since 1938 and the US since 1958, with no reported ill effects.Cry proteins selectively target Lepidopterans (moths and butterflies). As a toxic mechanism, cry proteins bind to specific receptors on the membranes of mid-gut (epithelial) cells, resulting in their rupture. Any organism that lacks the appropriate receptors in its gut is unaffected by the cry protein, and therefore is not affected by Bt. Regulatory agencies assess the potential for transgenic plants to affect non-target organisms before approving their commercial release.

In 1999, a paper stated that, in a laboratory environment, pollen from Bt maize dusted onto milkweed could harm the monarch butterfly. A collaborative research exercise over the following two years by several groups of scientists in the US and Canada studied the effects of Bt pollen in both the field and the laboratory. The study resulted in a risk assessment concluding that any risk posed to butterfly populations was negligible. A 2002 review of the scientific literature concluded that "the commercial large-scale cultivation of current Bt–maize hybrids did not pose a significant risk to the monarch population" and noted that despite large-scale planting of genetically modified crops, the butterfly's population was increasing. However, the herbicide glyphosate used to grow GMOs kills milkweed, the only food source of monarch butterflies, and by 2015 about 90% of the

U.S. population has declined. Lövei et al. analyzed laboratory settings and found that Bt toxins could affect non-target organisms, generally closely related to the intended targets. Typically, exposure occurs through the consumption of plant parts, such as pollen or plant debris, or through Bt ingestion by predators. A group of academic scientists criticized the analysis, writing: "We are deeply concerned about the inappropriate methods used in their paper, the lack of ecological context, and the authors' advocacy of how laboratory studies on non-target arthropods should be conducted and interpreted".

Biodiversity

Crop genetic diversity might decrease due to the development of superior GM strains that crowd others out of the market. Indirect effects might affect other organisms. To the extent that agrochemicals impact biodiversity, modifications that increase their use, either because successful strains require them or because the accompanying development of resistance will require increased amounts of chemicals to offset increased resistance in target organisms.

Studies comparing the genetic diversity of cotton found that in the US diversity has either increased or stayed the same, while in India it has declined. This difference was attributed to the larger number of modified varieties in the US compared to India.A review of the effects of Bt crops on soil ecosystems found that in general they "appear to have no consistent, significant, and long-term effects on the microbiota and their activities in soil".

The diversity and number of weed populations has been shown to decrease in farm-scale trials in the United Kingdom and in Denmark when comparing herbicide-resistant crops to their conventional counterparts. The UK trial suggested that the diversity of birds could be adversely affected by the decrease in weed seeds available for foraging. Published farm data involved in the trials showed that seed-eating birds were more abundant on conventional maize after the application of the herbicide, but that there were no significant differences in any other crop or prior to herbicide treatment. A 2012 study found a correlation between the reduction of milkweed in farms that grew glyphosate-resistant crops and the decline in adult monarch butterfly populations in Mexico.The New York Times reported that the study "raises the somewhat radical notion that perhaps weeds on farms should be protected. A 2005 study, designed to "simulate the impact of a direct overspray on a wetland" with four different agrochemicals (carbaryl (Sevin), malathion, 2,4-dichlorophenoxyacetic acid, and glyphosate in a Roundup formulation) by creating artificial ecosystems in tanks and then applying "each chemical at the manufacturer's maximum recommended application rates" found that "species richness was reduced by 15% with Sevin, 30% with malathion, and 22% with Roundup, whereas 2,4-D had no effect".The study has been used by environmental groups to argue that use of agrochemicals causes unintended harm to the environment and to biodiversity.

Secondary pests

Several studies documented surges in secondary pests within a few years of adoption of Bt cotton. In China, the main problem has been with mirids, which have in some cases "completely eroded all benefits from Bt cotton cultivation".A 2009 study in China concluded that the increase in secondary pests depended on local temperature and rainfall conditions and occurred in half the villages studied. The increase in insecticide use for the control of these secondary insects was far smaller than the reduction in total insecticide use due to Bt cotton adoption. A 2011 study based on a survey of 1,000 randomly selected farm households in five provinces in China found that the reduction in pesticide use in Bt cotton cultivars was significantly lower than that reported in research elsewhere: The finding was consistent with a hypothesis that more pesticide sprayings are needed over time to control emerging secondary pests, such as aphids, spider mites, and lygus bugs. Similar problems have been reported in India, with mealy bugs and aphids.

Gene flow

Genes from a GMO may pass to another organism just like an endogenous gene. The process is known as outcrossing and can occur in any new open-pollinated crop variety. Introduced traits potentially can cross into neighboring plants of the same or closely related species through three different types of gene flow: crop-to-crop, crop-to-weedy, and crop-to-wild. In crop-to-crop, genetic information from a genetically modified crop is transferred to a non-genetically modified crop. Crop-to-weedy transfer refers to the transfer of genetically modified material to a weed, and crop-to-wild indicates transfer from a genetically modified crop to a wild, undomesticated plant and/or crop. There are concerns that the spread of genes from modified organisms to unmodified relatives could produce species of weeds resistant to herbicides that could contaminate nearby non-genetically modified crops, or could disrupt the ecosystem, This is primarily a concern if the transgenic organism has a significant survival capacity and can increase in frequency and persist in natural populations.This process, whereby genes are transferred from GMOs to wild relatives, is different from the development of so-called "superweeds" or "superbugs" that develop resistance to pesticides under natural selection.

- In most countries environmental studies are required before approval of a GMO for commercial purposes, and a monitoring plan must be presented to identify unanticipated gene flow effects.
- In 2004, Chilcutt and Tabashnik found Bt protein in kernels of in a refuge (a conventional crop planted to harbor pests that might otherwise become resistant a pesticide associated with the GMO) implying that gene flow had occurred.
- In 2005, scientists at the UK Centre for Ecology and Hydrology reported the first evidence of horizontal gene transfer of pesticide resistance to weeds, in a few plants from a single season; they found no evidence that any of the hybrids had survived in subsequent seasons.

- In 2007, the U.S. Department of Agriculture fined Scotts Miracle-Gro $500,000 when modified DNA from GM creeping bentgrass, was found within relatives of the same genus (Agrostis) as well as in native grasses up to 21 km (13 mi) from the test sites, released when freshly cut, wind-blown grass.

In 2009, Mexico created a regulatory pathway for GM maize, but because Mexico is maize's center of diversity, concerns were raised about GM maize's effects on local strains. A 2001 report found Bt maize cross-breeding with conventional maize in Mexico.The data in this paper was later described as originating from an artifact and the publishing journal Nature stated that "the evidence available is not sufficient to justify the publication of the original paper", although it did not retract the paper. A subsequent large-scale study, in 2005, found no evidence of gene flow in Oaxaca. However, other authors claimed to have found evidence of such gene flow.

A 2010 study showed that about 83 percent of wild or weedy canola tested contained genetically modified herbicide resistance genes. According to the researchers, the lack of reports in the United States suggested that oversight and monitoring were inadequate.A 2010 report stated that the advent of glyphosate-resistant weeds could cause GM crops to lose their effectiveness unless farmers combined glyphosate with other weed-management strategies.

One way to avoid environmental contamination is genetic use restriction technology (GURT), also called "Terminator".This uncommercialized technology would allow the production of crops with sterile seeds, which would prevent the escape of GM traits. Groups concerned about food supplies had expressed concern that the technology would be used to limit access to fertile seeds. Another hypothetical technology known as "Traitor" or "T-GURT", would not render seeds sterile, but instead would require application of a chemical to GM crops to activate engineered traits. Groups such as Rural Advancement Foundation International raised concerns that further food safety and environmental testing needed to be done before T-GURT would be commercialized.

Escape of Modified Crops

- The escape of genetically modified seed into neighboring fields, and the mixing of harvested products, is of concern to farmers who sell to countries that do not allow GMO imports.
- In 1999 scientists in Thailand claimed they had discovered unapproved glyphosate-resistant GM wheat in a grain shipment, even though it was only grown in test plots. No mechanism for the escape was identified.
- In 2000, AventisStarLink GM corn was found in US markets and restaurants. It became the subject of a recall that started when Taco Bell-branded taco shells sold in supermarkets were found to contain it. StarLink was then discontinued.

Registration for Starlink varieties was voluntarily withdrawn by Aventis in October 2000.

- American rice exports to Europe were interrupted in 2006 when the LibertyLink modification was found in commercial rice crops, although it had not been approved for release.An investigation by the USDA's Animal and Plant Health Inspection Service (APHIS) failed to determine the cause of the contamination.

In May 2013, unapproved glyphosate-resistant GM wheat (but that had been approved for human consumption)was discovered in a farm in Oregon in a field that had been planted with winter wheat. The strain was developed by Monsanto, and had been field-tested from 1998 to 2005. The discovery threatened US wheat exports which totaled $8.1 billion in 2012. Japan, South Korea and Taiwan temporarily suspended winter wheat purchases as a result of the discovery. As of August 30, 2013, while the source of the modified wheat remained unknown, Japan, South Korea and Taiwan had resumed placing orders.

Coexistence with conventional crops

The US has no legislation governing the relationship among mixtures of farms that grow organic, conventional, and GM crops. The country relies on a "complex but relaxed" combination of three federal agencies (FDA, EPA, and USDA/APHIS) and states' common law tort systems to manage coexistence. The Secretary of Agriculture convened an Advisory Committee on Biotechnology and 21st Century Agriculture (AC21) to study coexistence and make recommendations about the issue. The members of AC21 included representatives of the biotechnology industry, the organic food industry, farming communities, the seed industry, food manufacturers, State governments, consumer and community development groups, the medical profession, and academic researchers. AC21 recommended that a study assess the potential for economic losses to US organic farmers; that any serious losses lead to a crop insurance program, an education program to ensure that organic farmers put appropriate contracts in place and that neighboring GMO farmers take appropriate containment measures. Overall the report supported a diverse agriculture system supporting diverse farming systems.

The EU implemented regulations specifically governing co-existence and traceability. Traceability has become commonplace in the food and feed supply chains of most countries, but GMO traceability is more challenging given strict legal thresholds for unwanted mixing. Since 2001, conventional and organic food and feedstuffs can contain up to 0.9% of authorised modified material without carrying a GMO label. (any trace of non-authorised modification is cause for a shipment to be rejected). Authorities require the ability to trace, detect and identify GMOs, and the several countries and interested parties created a non-governmental organization, Co-Extra, to develop such methods.

Chemical Use

Pesticides

Pesticides destroy, repel or mitigate pests (an organism that attacks or competes with a crop). A 2014 meta-analysis covering 147 original studies of farm surveys and field trials, and 15 studies from the researchers conducting the study, concluded that adoption of GM technology had reduced chemical pesticide use by 37%, with the effect larger for insect-tolerant crops than herbicide-tolerant crops.Some doubt still remains on whether the reduced amounts of pesticides used actually invoke a lower negative environmental effect, since there is also a shift in the types of pesticides used, and different pesticides have different environmental effects. In August 2015, protests occurred in Hawaii over the possibility that birth defects were being caused by the heavy use of pesticides on new strains of GM crops being developed there. Hawaii uses 17 times the amount of pesticides per acre compared to the rest of the US.

Herbicides

The development of glyphosate-tolerant (Roundup Ready) plants changed the herbicide use profile away from more persistent, higher toxicity herbicides, such as atrazine, metribuzin and alachlor, and reduced the volume and harm of herbicide runoff.A study by Chuck Benbrook concluded that the spread of glyphosate-resistant weeds had increased US herbicide use. That study cited a 23% increase (.3 kilograms/hectare) for soybeans from 1996–2006, a 43% (.9 kg/ha) increase for cotton from 1996–2010 and a 16% (.5 kg/ha) decrease for corn from 1996–2010. However, this study came under scrutiny because Benbrook did not consider the fact that glyphosate is less toxic than other herbicides, thus net toxicity may decrease even as use increases. Graham Brookes accused Benbrook of subjective herbicide estimates because his data, provided by the National Agricultural Statistics Service, does not distinguish between genetically modified and non-genetically modified crops. Brookes had earlier published a study that found that the use of biotech crops had reduced the volume and environmental impact of herbicide and other pesticides, which contradicted Benbrook. Brookes stated that Benbrook had made "biased and inaccurate" assumptions.

Insecticides

A claimed environmental benefit of Bt-cotton and maize is reduced insecticide use. A PG Economics study concluded that global pesticide use was reduced by 286,000 tons in 2006, decreasing pesticidal environmental impact by 15%.A survey of small Indian farms between 2002 and 2008 concluded that Bt cotton adoption had led to higher yields and lower pesticide use. Another study concluded that insecticide use on cotton and corn during the years 1996 to 2005 fell by 35,600,000 kilograms (78,500,000 lb) of active ingredient, roughly equal to the annual amount applied in the European Union. A Bt cotton study in six northern Chinese provinces from 1990 to 2010 concluded that it halved the use of pesticides and doubled the level of ladybirds, lacewings and spiders and extended environmental benefits to neighbouring crops of maize, peanuts and soybeans.

Resistant insect pests

Resistance evolves naturally after a population has been subjected to selection pressure via repeated use of a single pesticide. In November 2009, Monsanto scientists found that the pink bollworm had become resistant to first generation Bt cotton in parts of Gujarat, India—that generation expresses one Bt gene, Cry1Ac. This was the first instance of Bt resistance confirmed by Monsanto. Similar resistance was later identified in Australia, China, Spain and the US.

One strategy to delay Bt-resistance is to plant pest refuges using conventional crops, thereby diluting any resistant genes. Another is to develop crops with multiple Bt genes that target different receptors within the insect.In 2012, a Florida field trial demonstrated that army worms were resistant to Dupont-Dow's GM corn. This resistance was discovered in Puerto Rico in 2006, prompting Dow and DuPont to stop selling the product there. The European corn borer, one of Bt's primary targets, is also capable of developing resistance.

Economy

GM food's economic value to farmers is one of its major benefits, including in developing nations. A 2010 study found that Bt corn provided economic benefits of $6.9 billion over the previous 14 years in five Midwestern states. The majority ($4.3 billion) accrued to farmers producing non-Bt corn. This was attributed to European corn borer populations reduced by exposure to Bt corn, leaving fewer to attack conventional corn nearby. Agriculture economists calculated that "world surplus [increased by] $240.3 million for 1996. Of this total, the largest share (59%) went to U.S. farmers. Seed company Monsanto received the next largest share (21%), followed by US consumers (9%), the rest of the world (6%), and the germplasm supplier, Delta and Pine Land Company (5%)." PG Economics comprehensive 2012 study concluded that GM crops increased farm incomes worldwide by $14 billion in 2010, with over half this total going to farmers in developing countries.

The main Bt crop grown by small farmers in developing countries is cotton. A 2006 review of Bt cotton findings by agricultural economists concluded, "the overall balance sheet, though promising, is mixed. Economic returns are highly variable over years, farm type, and geographical location". However, environmental activist Mark Lynas said that complete rejection of genetic engineering is "illogical and potentially harmful to the interests of poorer peoples and the environment".

In 2013, the European Academies Science Advisory Council (EASAC) asked the EU to allow the development of agricultural GM technologies to enable more sustainable agriculture, by employing fewer land, water and nutrient resources. EASAC also criticizes the EU's "timeconsuming and expensive regulatory framework" and said that the EU had fallen behind in the adoption of GM technologies.

Industrial agriculture

GM crops play a key role in intensive crop farming, which involves monoculture, use of herbicides and pesticides, use of equipment requiring large amounts of fuel and irrigation. Opponents such as Jonathan Latham of the Bioscience Resource Project and Vandana Shiva treat industrial agriculture and modified crops as closely related topics, and call for agriculture to adopt practices drastically reducing disruptions to the environment. Proponents of conventional agriculture point to its high yields, low prices, and wider choices and claim that technology is necessary to feed a growing world population.

Developing nations

Disagreements about developing nations include the claimed need for increased food supplies, and how to achieve such an increase. Some scientists suggest that a second Green Revolution including use of modified crops is needed to provide sufficient food. The potential for genetically modified food to help developing nations was recognised by the International Assessment of Agricultural Science and Technology for Development, but as of 2008 they had found no conclusive evidence of a solution. Skeptics such as John Avise claim that apparent shortages are caused by problems in food distribution and politics, rather than production. Other critics say that the world has so many people because the second green revolution adopted unsustainable agricultural practices that left the world with more mouths to feed than the planet can sustain. Pfeiffer claimed that even if technological farming could feed the current population, its dependence on fossil fuels, which in 2006 he incorrectly predicted would reach peak output in 2010, would lead to a catastrophic rise in energy and food prices.

Claimed deployment constraints to developing nations include the lack of easy access, equipment costs and intellectual property rights that hurt developing countries. The Consultative Group on International Agricultural Research (CGIAR), an aid and research organization, was praised by the World Bank for its efforts, but the bank recommended that they shift to genetics research and productivity enhancement.

Obstacles include access to patents, commercial licenses and the difficulty that developing countries have in accessing genetic resources and other intellectual property. The International Treaty on Plant Genetic Resources for Food and Agriculture attempted to remedy this problem, but results have been inconsistent. As a result, "orphan crops", such as teff, millets, cowpeas and indigenous plants, which are important in these countries receive little investment. Writing about Norman Borlaug's 2000 publication Ending world hunger: the promise of biotechnology and the threat of antiscience zealotry, the authors argued that Borlaug's warnings were still true in 2010: GM crops are as natural and safe as today's bread wheat, opined Dr. Borlaug, who also reminded agricultural scientists of their moral obligation to stand up to the antiscience crowd and warn policy makers that global food insecurity will not disappear without this new technology and ignoring this reality would make future solutions all the more difficult to achieve.

Yield

US maize yields were flat until the 1930s, when the adoption of conventional hybrid seeds caused them to increase by ~.8 bushels/acre (1937–1955). Thereafter a combination of improved genetics, fertilizer and pesticide availability and mechanization raised the rate of increase to 1.9 bushels per acre per year. In the years since the advent of GM maize, the rate increased slightly to 2.0. Average US maize yields were 174.2 bushels per acre in 2014.

Commercial GM crops have traits that reduce yield loss from insect pressure or weed interference.

2014 Review

A 2014 review, concluded that GM crops' effects on farming were positive. According to The Economist, the meta-analysis considered all published English-language examinations of the agronomic and economic impacts between 1995 and March 2014. The study found that herbicide-tolerant crops have lower production costs, while for insect-resistant crops the reduced pesticide use was offset by higher seed prices, leaving overall production costs about the same.

Yields increased 9% for herbicide tolerance and 25% for insect resistance. Farmers who adopted GM crops made 69% higher profits than those who did not. The review found that GM crops help farmers in developing countries, increasing yields by 14 percentage points. The researchers considered some studies that were not peer-reviewed, and a few that did not report sample sizes. They attempted to correct for publication bias, by considering sources beyond academic journals. The large data set allowed the study to control for potentially confounding variables such as fertiliser use. Separately, they concluded that the funding source did not influence study results.

2010 Review

A 2010 article, supported by CropLife International summarised the results of 49 peer reviewed studies. On average, farmers in developed countries increased yields by 6% and 29% in developing countries. Tillage decreased by 25–58% on herbicide-resistant soybeans. Glyphosate-resistant crops allowed farmers to plant rows closer together as they did not have to control post-emergent weeds with mechanical tillage. Insecticide applications on Bt crops were reduced by 14–76%. 72% of farmers worldwide experienced positive economic results.

2009 Review

In 2009, the Union of Concerned Scientists, a group opposed to genetic engineering and cloning of food animals, summarized peer-reviewed studies on the yield contribution of GM soybeans and maize in the US.The report concluded that other agricultural methods had made a greater contribution to national crop yield increases in recent years than genetic engineering.

Wisconsin study

A study unusually published as correspondence rather than as an article examined maize modified to express four traits (resistance to European corn borer, resistance to corn root worm, glyphosate tolerance and glyfosinate tolerance) singly and in combination in Wisconsin fields from 1990–2010.The variance in yield from year to year was reduced, equivalent to a yield increase of 0.8–4.2 bushels per acre. Bushel per acre yield changes were +6.4 for European corn borer resistance, +5.76 for glufosinate tolerance, "5.98 for glyphosate tolerance and "12.22 for corn rootworm resistance. The study found interactions among the genes in multi-trait hybrid strains, such that the net effect varied from the sum of the individual effects. For example, the combination of European corn borer resistance and glufosinate tolerance increased yields by 3.13, smaller than either of the individual traits

Market dynamics

The seed industry is dominated by a small number of vertically integrated firms. In 2011, 73% of the global market was controlled by 10 companies.

In 2001, the USDA reported that industry consolidation led to economies of scale, but noted that the move by some companies to divest their seed operations questioned the long-term viability of these conglomerates. Two economists have said that the seed companies' market power could raise welfare despite their pricing strategies, because "even though price discrimination is often considered to be an unwanted market distortion, it may increase total welfare by increasing total output and by making goods available to markets where they would not appear otherwise."

Market share gives firms the ability to set or influence price, dictate terms, and act as a barrier to entry. It also gives firms bargaining power over governments in policy making. In March 2010, the US Department of Justice and the US Department of Agriculture held a meeting in Ankeny, Iowa, to look at the competitive dynamics in the seed industry. Christine Varney, who heads the antitrust division in the Justice Department, said that her team was investigating whether biotech-seed patents were being abused.A key issue was how Monsanto licenses its patented glyphosate-tolerance trait that was in 93 percent of US soybeans grown in 2009. About 250 family farmers, consumers and other critics of corporate agriculture held a town meeting prior to the government meeting to protest Monsanto's purchase of independent seed companies, patenting seeds and then raising seed prices.

Intellectual property

Traditionally, farmers in all nations saved their own seed from year to year. However, since the early 1900s hybrid crops have been widely used in the developed world and seeds to grow these crops are purchased each year from seed producers.The offspring of the hybrid corn, while still viable, lose hybrid vigor (the beneficial traits of the parents). This benefit of first-generation hybrid seeds is the primary reason for not planting second-generation seed. However, for non-hybrid GM crops, such as GM soybeans, seed companies use intellectual property law and tangible property common law, each expressed in contracts, to prevent farmers from planting saved seed. For example, Monsanto's typical bailment license (covering transfer of the seeds themselves) forbids saving seeds, and also requires purchasers to sign a separate patent license agreement.

Corporations say that they need to prevent seed piracy, to fulfill financial obligations to shareholders, and to finance further development. DuPont spent approximately half its $2 billion research and development (R&D) budget on agriculture in 2011 while Monsanto spends 9–10% of sales on R&D.

Detractors such as Greenpeace say that patent rights give corporations excessive control over agriculture. The Center for Ecoliteracy claimed that "patenting seeds gives companies excessive power over something that is vital for everyone". A 2000 report stated, "If the rights to these tools are strongly and universally enforced - and not extensively licensed or provided pro bono in the developing world - then the potential applications of GM technologies described previously are unlikely to benefit the less developed nations of the world for a long time" (i.e. until after the restrictions expire). Monsanto has patented its seed and it obligates farmers who choose to buy its seeds to sign a license agreement, obligating them store or sell, but not plant, all the crops that they grow.

Besides large agri-businesses, in some instances, GM crops are also provided by science departments or research organisations which have no commercial interests.

Lawsuits filed against farmers for patent infringement

Monsanto has filed patent infringement suits against 145 farmers, but proceeded to trial with only 11.In some of the latter, the defendants claimed unintentional contamination by gene flow, but Monsanto won every case. Monsanto Canada's Director of Public Affairs stated, "It is not, nor has it ever been Monsanto Canada's policy to enforce its patent on Roundup Ready crops when they are present on a farmer's field by accident ... Only when there has been a knowing and deliberate violation of its patent rights will Monsanto act." In 2009 Monsanto announced that after its soybean patent expires in 2014, it will no longer prohibit farmers from planting soybean seeds that they grow.

One example of such litigation is the Monsanto v. Schmeiser case.This case is widely misunderstood.In 1997, Percy Schmeiser, a canola breeder and grower in Bruno, Saskatchewan, discovered that one of his fields had canola that was resistant to Roundup.

He had not purchased this seed, which had blown onto his land from neighboring fields. He later harvested the area and saved the crop in the back of a pickup truck. Before the 1998 planting, Monsanto representatives informed Schmeiser that using this crop for seed would infringe the patent, and offered him a license, which Schmeiser refused. According to the Canadian Supreme Court, after this conversation "Schmeiser nevertheless took the harvest he had saved in the pick-up truck to a seed treatment plant and had it treated for use as seed. Once treated, it could be put to no other use. Mr. Schmeiser planted the treated seed in nine fields, covering approximately 1,000 acres in all ... A series of independent tests by different experts confirmed that the canola Mr. Schmeiser planted and grew in 1998 was 95 to 98 percent Roundup resistant." After further negotiations between Schmeiser and Monsanto broke down, Monsanto sued Schmeiser for patent infringement and prevailed in the initial case. Schmeiser appealed and lost, and appealed again to the Canadian Supreme Court, which in 2004 ruled 5 to 4 in Monsanto's favor, stating that "it is clear on the findings of the trial judge that the appellants saved, planted, harvested and sold the crop from plants containing the gene and plant cell patented by Monsanto".

International trade

GM crops have been the source of international trade disputes and tensions within food-exporting nations over whether introduction of genetically modified crops would endanger exports to other countries.

In Canada in 2010, flax exports to Europe were rejected when traces of an experimental GM flax were found in shipments.This led a member of Parliament to propose Private Member's Bill C-474, which would have required that "an analysis of potential harm to export markets be conducted before the sale of any new genetically engineered seed is permitted". Opponents claimed that "incorporating stringent socio-economic standards into the science-based regulatory system could spell the end of private research funding; because if private biotechnology companies can't see the possibility of a return on their investment, they'll invest their research budget elsewhere".The bill was defeated 176 to 97 in 2011.

Regulation

Labeling

Green: Mandatory labeling required; Red: Ban on import and cultivation of genetically engineered food. Regulations are pending in Israel and the United States.

In 2014, 64 countries required labeling of all GM foods. These include the European Union,Japan,Australia,New Zealand,Russia,China and India. As of March 2015, Israel was in the process of issuing regulations for labeling of food with ingredients from GMOs.

Alaska required labeling of GMO fish and shellfish in 2005, even though no GM fish had been approved by the FDA at the time. A 2014 Vermont law went into effect on July 1, 2016, and some food manufacturers (including General Mills, Mars, Kellogg's, the Campbell Soup Company, PepsiCo, ConAgra, Frito-Lay, and Bimbo Bakeries USA) began distributing products either locally or nationwide with labels such as "Partially produced with Genetic Engineering". Other manufacturers removed about 3,000 non-compliant products from sale in Vermont. The federal government of the United States passed a law at the end of that month pre-empting all state laws, including Vermont's. The law requires labeling regulations to be issued by July 2018, and allows indirect disclosure such as with a phone number, bar code, or web site. It is unclear whether the rules will require labeling of oils and sugars from GM crops, where the final product does not contain any "genetic material" as mentioned in the law.

Prior to the new federal rules taking effect, while it does require pre-market approval, the U.S. Food and Drug Administration has not required GMO labeling as long as there are no differences in health, environmental safety, and consumer expectations based on the packaging. The federal rules come after GMO labeling was debated in many state legislatures and defeated in popular referendums in Oregon (2002 and 2014), Colorado (2014),California Proposition 37 (2012), and Washington Initiative 522 (2012). Connecticut and Maine had passed laws in 2013 and 2014 respectively, which would have required GMO food labels if Northeast states with a population of at least 20 million had passed similar laws (and for Connecticut, representing at least four states).

Other jurisdictions make such labeling voluntary or have had plans to require labeling. Major GM food crop exporters like the United States (until 2018), Argentina, and Canada have adopted voluntary labeling approaches; China and Brazil have major GM (largely non-food) crops and have adopted mandatory labelling.

Arguments

The American Public Health Association, the British Medical Association and the Public Health Association of Australia support mandatory labeling. The European Commission argued that mandatory labeling and traceability are needed to allow for informed choice, avoid potential misleading of consumers and facilitate the withdrawal of products if adverse effects on health or the environment are discovered.A 2007 study on the effect of labeling laws found that once labeling went into effect, few products continued to contain GM ingredients. The study also found that costs were higher in food-exporting than in food-importing countries.

The American Medical Association (AMA) and the American Association for the Advancement of Science have opposed mandatory labeling absent scientific evidence of harm. The AMA said that even voluntary labeling is misleading unless accompanied by focused consumer education. The AAAS stated that mandatory labeling "can only serve to mislead and falsely alarm consumers".

[Labeling] efforts are not driven by evidence that GM foods are actually dangerous. Indeed, the science is quite clear: crop improvement by the modern molecular techniques of biotechnology is safe. Rather, these initiatives are driven by a variety of factors, ranging from the persistent perception that such foods are somehow "unnatural" and potentially dangerous to the desire to gain competitive advantage by legislating attachment of a label meant to alarm. Another misconception used as a rationale for labeling is that GM crops are untested.

Objectivity of regulatory bodies

Groups such as the Union of Concerned Scientists and Center for Food Safety that have expressed concerns about the FDA's lack of a requirement for additional testing for GMO's, lack of required labeling and the presumption that GMO's are "Generally Recognized as Safe" (GRAS), have questioned whether the FDA is too close to companies that seek approval for their products.

Critics in the U.S. protested the appointment of lobbyists to senior positions in the Food and Drug Administration. Michael R. Taylor, a former Monsanto lobbyist, was appointed as a senior adviser to the FDA on food safety in 1991. After leaving the FDA, Taylor became a vice-president of Monsanto. On 7 July 2009, Taylor returned to government as a senior adviser to the FDA Commissioner.

In 2001, when the Starlink corn recall became public, the U.S. Environmental Protection Agency was criticized for being slow to react by Joseph Mendelson III of the Center for Food Safety. He also criticized the EPA and Aventis CropScience for statements at the time of the recall, that indicated they did not anticipate that such a thing would happen. The Canadian Biotechnology Advisory Committee that reviewed Canada's regulations in 2003 was accused by environmental and citizen groups of not representing the full spectrum of public interests and for being too closely aligned to industry groups.

Most of the Chinese National Biosafety Committee are involved in biotechnology, a situation that led to criticisms that they do not represent a wide enough range of public concerns.

Litigation and Regulation Disputes

United States

Four federal district courtsuits have been brought against Animal and Plant Health Inspection Service (APHIS), the agency within USDA that regulates genetically modified plants. Two involved field trials (herbicide-tolerant turfgrass in Oregon; pharmaceutical-producing corn and sugar in Hawaii) and two the deregulation of GM alfalfa. and GM sugar beet. APHIS lost all four cases at trial, with the judges ruling they failed to diligently follow the guidelines set out in the National Environmental Policy Act. However, the Supreme Court overturned the nationwide ban on GM alfalfa and an appeal court allowed the partial deregulation of GM sugar beets.After APHIS prepared Environmental Impact Statements for both alfalfa and sugar beets they were approved.

In 2014, Maui County, Hawaii approved an initiative calling for a moratorium on GMO production and research. The initiative specified penalties including fines and jail for knowing violations and did not limit its scope to commercial agriculture. The initiative passed by about 50.2 to 47.9 percent. On December 15, 2015, the New York Times ran an op-ed titled "Are You Eating Frankenfish?", saying that the United States congress will debate whether genetically engineered salmon should be labeled.

European Union

Until the 1990s, Europe's regulation was less strict than in the U.S. In 1998, the use of MON810, a Bt expressing maize conferring resistance to the European corn borer, was approved for commercial cultivation in Europe. However, in the 1990s a series of unrelated food crises created consumer apprehension about food safety in general and eroded public trust in government oversight. A bovine spongiform encephalopathy outbreak was the most publicized.In 1998, a de facto moratorium led to the suspension of approvals of new GMOs in the EU pending the adoption of revised rules. In the mid-1990s, government approval of some GMO crops in the United States precipitated public concern in Europe and led to a dramatic decrease in American exports to Europe. "Prior to 1997, corn exports to Europe represented about 4% of total US corn exports, generating about $300 million in sales ... For example, before 1997, the U.S. sold about 1.75 million tons of corn annually to Spain and Portugal ... But in the 1998–99 crop year, Spain bought less than a tenth of the previous year's amount and Portugal bought none at all."

In May 2003, the US and twelve other countries filed a formal complaint with the World Trade Organization that the EU was violating international trade agreements, by blocking imports of US farm products through its ban on GM food. The countries argued that the EU's regulatory process was far too slow and its standards were unreasonable given the scientific evidence showing that the crops were safe. The case was lobbied by Monsanto and France's Aventis, as well as by US agricultural groups such as the National Corn Growers Association. In response, in June 2003, the European Parliament ratified a U.N.biosafety protocol regulating international trade in GM food, and in July agreed to new regulations requiring labeling and traceability, as well as an opt-out provision for individual countries. The approval of new GMOs resumed in May 2004. While GMOs have been approved since then, approvals remain controversial and various countries have utilized opt-out provisions. In 2006, the World Trade Organization ruled that the pre-2004 restrictions had been violations, although the ruling had little immediate effect since the moratorium had already been lifted. In late 2007, the US ambassador to France recommended "moving to retaliation" to cause "some pain" against France and the European Union in an attempt to fight the French ban and changes in European policy toward genetically modified crops, according to a US government diplomatic cable obtained by WikiLeaks. 20 out of 28 European Countries (including Switzerland) said No to GMOs until October 2015.

Australia

In May 2014, the Supreme Court of the Australian state of Western Australia dismissed "Marsh v. Baxter". The plaintiff was Steve Marsh, an organic farmer, and the defendant was Michael Baxter, his lifelong neighbour, who grew GM canola.In late 2010, Marsh found seeds from Baxter's crop in his fields. Later, Marsh found escaped GM canola growing amidst his crop. Marsh reported the seed and plants to his local organic certification board, and lost the organic certification of some 70 per cent of his 478 hectare farm. Marsh sued on the grounds that Baxter used a method of harvesting his crop that was substandard and negligent, and on the basis that his land had been widely contaminated. In its summary judgment, the court found that approximately 245 cut canola plants were blown by the wind into Marsh's property, Eagle's Rest. However, Baxter's method (swathing) was "orthodox and well accepted harvest methodology". "In 2011, eight GM canola plants were found to have grown up as self-sown volunteer plants on Eagle Rest", which "were identified and pulled out", and "no more volunteer RR canola plants grew on Eagle Rest in subsequent years". The summary judgment stated that the loss of organic certification "was occasioned by the erroneous application of governing NASAA Standards applicable to NASAA organic operators as regards GMOs (genetically modified organisms) at the time". and that "[t]he absence of a reliable underlying evidentiary platform to support a perpetual injunction against swathing was a significant deficiency".

On June 18, 2014, Marsh announced that he had filed an appeal. One ground was the costs of $803,989 awarded against him. The appeal hearing commenced on 23 March 2015 and was adjourned on 25 March "to deal with an order to ascertain whether Mr Baxter's defence has been financially supported by GM-seed supplier Monsanto and/or the Pastoralists and Graziers Association (PGA)". The Court of Appeal subsequently dismissed the appeal and ordered Marsh to pay Baxter's costs.

Philippines

A petition filed May 17, 2013, by environmental group Greenpeace Southeast Asia and farmer-scientist coalition Masipag (Magsasaka at Siyentipiko sa Pagpapaunlad ng Agrikultura) asked the appellate court to stop the planting of Bt eggplant in test fields, saying the impacts of such an undertaking to the environment, native crops and human health are still unknown.

The Court of Appeals granted the petition, citing the precautionary principle stating "when human activities may lead to threats of serious and irreversible damage to the environment that is scientifically plausible but uncertain, actions shall be taken to avoid or diminish the threat". Respondents filed a motion for reconsideration in June 2013 and on September 20, 2013 the Court of Appeals chose to uphold their

May decision saying the bt talong field trials violate the people's constitutional right to a "balanced and healthful ecology". The Supreme Court on December 8, 2015, permanently stopped the field testing for Bt (Bacillus thuringiensis) talong (eggplant), upholding the decision of the Court of Appeals which stopped the field trials for the genetically modified eggplant.

Innovation in Technology and Regulatory law

The first genetically modified crops were made with transgenic approaches, introducing foreign genes and sometimes using bacteria to transfer the genes. In the US, these foreign genetic elements placed the resulting plant under the jurisdiction of the USDA under the Plant Protection Act. However, as of 2010, newer genetic engineering technologies like genome editing have allowed scientists to modify plant genomes without adding foreign genes, thus escaping USDA regulation. Critics have called for regulation to be changed to keep up with changing technology.

Legislation

African Controversies

In 2002, in the midst of a famine, Zambia refused emergency food aid that contained food from genetically modified crops, based on the precautionary principle.

During a conference in the Ethiopian capital of Addis Ababa, Kingsley Amoako, Executive Secretary of the United Nations Economic Commission for Africa (UNECA), encouraged African nations to accept GM food and expressed dissatisfaction in the public's negative opinion of biotechnology.

Studies for Uganda showed that transgenic bananas had a high potential to reduce rural poverty but that urban consumers with a relatively higher income might reject them. Critics claimed that shipment of US food to southern Africa was more about promoting the adoption of biotech crops in the region than about hunger. The US was supplying Africa with meals and support during a food crisis they were facing in the early 2000s. However, once some of the African countries realized that these shipments contained GM maize, they rejected the shipments and stopped releasing the food that had been sent to them. Critics accused the US of "exploiting the Southern African famine as a public relations tool". The U.S. countered these comments by saying that European nations were letting millions of Africans suffer from hunger and starvation because of "irrational fears over hypothetical and unproven risks". The US had a pre-GMO policy of shipping US crops as food aid, rather than buying crops in/near the countries that needed aid. The US policy was claimed to be more costly than Europe's.

Indian controversies

India is an agrarian country with around 60% of its people depending directly or indirectly upon agriculture. From 1995 to 2013, a total of 296,438 farmers have killed themselves in India, or an average of 16,469 suicides per year. During the same period, about 9.5 million people died per year in India from other causes including malnutrition, diseases and suicides that were non-farming related, or about 171 million deaths from 1995 to 2013. Activists and scholars have offered a number of conflicting reasons for farmer suicides, such as monsoon failure, high debt burdens, genetically modified crops, government policies, public mental health, personal issues and family problems. There are also accusations of states reporting inaccurate data on farmer suicides.

In India, GM cotton yields in Maharashtra, Karnataka, and Tamil Nadu resulted in an average 42% increase in yield in 2002, the first year of commercial planting. A severe drought in Andhra Pradesh that year prevented any increase in yield, because the GM strain was not drought tolerant. Drought-tolerant variants were later developed. Driven by substantially reduced losses to insect predation, by 2011 88% of Indian cotton was modified.There are economic and environmental benefits of GM cotton to farmers in India. A study from 2002 through 2008 on the economic impacts of Bt cotton in India, showed that Bt cotton increased yields, profits and living standards of smallholder farmers. However, recently cotton bollworm has been developing resistance to Bt cotton. Consequently, in 2012 Maharashtra banned Bt cotton and ordered an independent socioeconomic study of its use. Indian regulators cleared the Bt brinjal, a genetically modified eggplant, for commercialisation in October 2009. After opposition by some scientists, farmers and environmental groups, a moratorium was imposed on its release in February 2010 "for as long as it is needed to establish public trust and confidence".

As of 1 January 2013, all foods containing GMOs must be labelled. The Legal Metrology (Packaged Commodities) Rules, 2011 states that "every package containing the genetically modified food shall bear at the top of its principal display panel the letters 'GM.'" The rules apply to 19 products including biscuits, breads, cereals and pulses, and a few others. The law faced criticism from consumer rights activists as well as from the packaged-food industry; both sides had major concerns that no logistical framework or regulations had been established to guide the law's implementation and enforcement. On March 21, 2014, the Indian government revalidated 10 GM-based food crops and allowed field trials of GM food crops, including wheat, rice and maize. Food sovereignty", a term coined by members of Via Campesina in 1996, asserts that the people who produce, distribute, and consume food should control the mechanisms and policies of food production and distribution, rather than the corporations and market institutions they believe have come to dominate the global food system. It also encompasses the right of peoples to healthy and culturally appropriate food and their right to define their own food and agriculture systems. The phrase "culturally appropriate" signifies that the food that is available and accessible for the population should fit with the cultural background of the people consuming it.

History

The history of food sovereignty as a movement is relatively young. However, there are a number of key movements and countries that have made significant steps towards making an alternative food system a reality.

Global gatherings

At the Forum for Food Sovereignty in Sélingué, Mali, 27 February 2007, about 500 delegates from more than 80 countries adopted the "Declaration of Nyéléni", which says in part: Food sovereignty is the right of peoples to healthy and culturally appropriate food produced through ecologically sound and sustainable methods, and their right to define their own food and agriculture systems. It puts those who produce, distribute and consume food at the heart of food systems and policies rather than the demands of markets and corporations. It defends the interests and inclusion of the next generation. It offers a strategy to resist and dismantle the current corporate trade and food regime, and directions for food, farming, pastoral and fisheries systems determined by local producers. Food sovereignty prioritises local and national economies and markets and empowers peasant and family farmer-driven agriculture, artisanal fishing, pastoralist-led grazing, and food production, distribution and consumption based on environmental, social and economic sustainability. In April 2008 the International Assessment of Agricultural Science and Technology for Development (IAASTD), an intergovernmental panel under the sponsorship of the United Nations and the World Bank, adopted the following definition: "Food sovereignty is defined as the right of peoples and sovereign states to democratically determine their own agricultural and food policies."

Becoming part of government policy

In September 2008, Ecuador became the first country to enshrine food sovereignty in its constitution. As of late 2008, a law is in the draft stages that is expected to expand upon this constitutional provision by banning genetically modified organisms, protecting many areas of the country from extraction of non-renewable resources, and to discourage monoculture. The law as drafted will also protect biodiversity as collective intellectual property and recognize the Rights of Nature.

Since then another five countries have integrated food sovereignty into their national constitutions or laws. These countries are Venezuela, Mali, Bolivia, Nepal and Senegal; and most recently Egypt (2014 Constitution).

BIBLIOGRAPHY

- Aanchal Sharma; Binod Kumar Singh; Anand, Neeraj (2016). Fruit Processing Industry in India: A Short Review. Cold Chain Logistics in Horticulture & Agriculture. Winsar Publishing Company. doi:10.13140/RG.2.1.2155.3047/1.
- American Community Gardening Association (1995). Monograph on the 1994 National Gardening Association Survey. Philadelphia.
- Armstrong, Wayne P. Identification of Major Fruit Types. Wayne's Word: An On-Line Textbook of Natural History. Archived from the original on 20 November 2011. Retrieved 17 August 2013.
- Ausherman, J; Ubbes, V; Kowalski, J (2014). Using School Gardening as a Vehicle for Critical and Creative Thinking in Health Education. Health Educator. pp. 41–48.
- Bassett, Thomas J. (1981). Reaping on the Margins: A Century of Community Gardening in America. 25 (2): 1–8.
- Beck, Margery A. Aronia berry gaining market foothold in U.S. USA Today. Retrieved 13 August 2015.
- Blueberries – Celebrating 100 Years. Blueberry Council. Retrieved 11 August 2015.
- Bring on the Blueberry Wine. Wine Mag. 19 September 2012. Retrieved 13 August 2015.
- DeEll, Dr. Jennifer. Postharvest Handling and Storage of Berries. omafra. Retrieved 12 August 2015.
- Fresh Fruit vs Frozen Fruit in baking recipes. Baking Bites. 26 July 2012. Retrieved 13 August 2015.
- Fruit Juices. Agriculture and Agri-Food Canada. Archived from the original on 4 December 2014. Retrieved 13 August 2015.

- Gao, Gary. Strawberries are an Excellent Fruit for the Home Garden. Ohio State University Extension. Retrieved 13 August 2015.
- Geisler, Malinda. Cranberries Profile. AgMRC. Retrieved 13 August 2015.
- Global Horticulture (2014 -2018) - Pink and Healthy. PRNewswire. 19 August 2014.
- Global Horticulture (2014 -2018) - Pink and Healthy. PRNewswire. 19 August 2014.
- Gordon, Megan. Frozen Berries In Off-Season Baking: Should You Thaw Before Using?. The Kitchn. Retrieved 13 August 2015.
- Growing greener cities in Africa (PDF). Rome: FAO. 2012. ISBN 978-92-5-107286-8.
- Growing Strawberries. University of Illinois Extension. Retrieved 13 August 2015.
- Handley, David T. Growing Strawberries. University of Maine Extension. Retrieved 13 August 2015.
- Honeyman, M (1992). Vegetation and stress: A comparison study of varying amounts of vegetation in countryside and urban scenes. Portland: In: The Role of Horticulture in Human Well-Being and Social Development: A National Symposium. Timber Press. pp. 143–145.
- Hynes, H.P; Howe, G (2004). Urban Horticulture in the Contemporary United States: Personal and Community Benefits. Acta Hort. (ISHS) 643. pp. 171–181.
- Hynes, Patricia (May 1, 1996). A Patch of Eden: America's Inner-City Gardeners (First ed.). Vermont: Chelsea Green Publishing Company.
- Jacobs, Jane (January 1, 1969). The Economy of Cities. New York: Random House.
- Kaplan, R; Kaplan, S (1990). Restorative experience: The healing power of nearby nature. Cambridge, Massachusetts: In: Francis and R.T. Hoster (eds). pp. 238–244.
- Karina Fernandez-Stark; Penny Bamber; Gary Gereffi (2011). The Fruit and Vegetables Global Value Chain – ECONOMIC UPGRADING AND WORKFORCE DEVELOPMENT (PDF). CGGC, Duke University. Archived from the original (PDF) on 16 September 2012. Retrieved 12 August 2016.
- Kevin M. Folta; Chittaranjan Kole (16 May 2011). Genetics, Genomics and Breeding of Berries. CRC Press. ISBN 978-1-4398-5660-4.
- Kime, Robert. Strawberry Wine (PDF). Berry Resources Cornell. Retrieved 13 August 2015.
- Kluepfel, Marjan; Polomski, Bob. Growing Strawberries. Clemson Cooperative Extension. Retrieved 13 August 2015.

- Lake, Barbara, Taciano L. Milfront C. Gavin. The Relative Influence of Psycho-Social Factors on Urban Gardening. Academic Search Premier. Web. 10 June 2013.
- Lea, Andrew G. H.; Piggott, John Raymond; Piggott, John R. (30 June 2003). Fermented Beverage Production. Springer Science & Business Media. ISBN 978-0-306-47706-5.
- Pittaway, Jenna (10 April 2014). Dr Barney Interview on the Western Huckleberry. wildhuckleberry. Retrieved 11 August 2015.
- Pritts, Dr. Marvin. Key Features of Organic Berry Crop Production (PDF). Cornell Fruit. Retrieved 11 August 2015.
- Pritts, Dr. Marvin. Site and Soil requirements for small fruit crops (PDF). Cornell Fruit. Retrieved 11 August 2015.
- Relf, Diane. Human Issues In Horticulture. Virginia Tech. Department of Horticulture. Retrieved 6 October 2014.
- Robinson, Muriel. Urban Horticulture- Purpose and Prospects. The Robinson Garden at Earlscliffe, Baily, Co. Dublin, Ireland. Retrieved 6 November 2014.
- Russell, Betsy Z. Wild huckleberry nearly tamed. idahoptv. Archived from the original on 3 August 2018. Retrieved 11 August 2015.
- Ruttan, Denise. Plant strawberries and boost your health. Oregon State University Extension Service. Archived from the original on 28 September 2015. Retrieved 13 August 2015.
- Strawberry Production Systems. Maine Organic Farmers and Gardners Association. Archived from the original on 6 September 2015. Retrieved 13 August 2015.
- Thiem B (2003). Rubus chamaemorus L. – a boreal plant rich in biologically active metabolites: a review (PDF). Biological Letters. 40 (1): 3–13. Archived from the original (PDF) on 9 August 2017. Retrieved 17 September 2020.
- Tukey, HB Jr. (1983). Urban horticulture: horticulture for populated areas. HortScience: 11–13.
- Whiting, David. Growing Strawberries in Colorado Gardens. Colorado State University Extension. Archived from the original on 21 August 2015.
- Wright, John (31 August 2011). How to make Blackberry Wine and Whisky. The Guardian. Retrieved 13 August 2015.
- Yanyun Zhao (6 June 2007). Berry Fruit: Value-Added Products for Health Promotion. CRC Press. ISBN 978-1-4200-0614-8.

- Zick, C; et al. (2013). Harvesting More Than Vegetables: The Potential Weight Control Benefits of Community Gardening. American Journal of Public Health. pp. 1110–1115.

INDEX

9 789392 969317

Printed at Repro India Ltd.